5th Conference on Constraint-Based Reconstruction and Analysis (COBRA 2018)

Seattle, Washington, USA
14 - 16 October 2018

ISBN: 978-1-5108-7587-6

Printed from e-media with permission by:

Curran Associates, Inc.
57 Morehouse Lane
Red Hook, NY 12571

Some format issues inherent in the e-media version may also appear in this print version.

Printed by Curran Associates, Inc. (2019)

AIChE
120 Wall Street, FL 23
New York, NY 10005-4020

Phone: (800) 242-4363
Fax: (203) 775-5177

www.aiche.org

Additional copies of this publication are available from:

Curran Associates, Inc.
57 Morehouse Lane
Red Hook, NY 12571 USA
Phone: 845-758-0400
Fax: 845-758-2633
Email: curran@proceedings.com
Web: www.proceedings.com

TABLE OF CONTENTS

GENERAL

POSTERS

Kinetic Modelling within Our GRASP: New Strategies for Fitting Accurate Network Kinetic Models

Authors:

Nielsen, L. K., Technical University of Denmark

Saa, P. A., Pontificia Universidad Católica de Chile

Several genome-scale metabolic reconstruction software platforms have been developed and continuously updated during the last fifteen years. These tools have been widely applied to reconstruct metabolic models for hundreds of microorganisms ranging from important human pathogens to species of industrial relevance. So far, these platforms have not been systematically evaluated with respect to software quality, best potential uses and intrinsic capacity to generate high-quality genome-scale metabolic models. In consequence, this lack of assessment keeps users away from selecting the tool that best fits the purpose of their research. In this work, we reviewed the current genome-scale reconstruction software platforms and we performed a systematic assessment of the most promising tools. We defined a list of features for assessing software quality related to genome-scale reconstruction which we hope it would be useful for potential users. In addition, we compared the output networks generated using each of these tools with the high-quality manually curated models of*Lactobacillus plantarum* and *Bortedella pertussis*, representatives of gram-positive and gram-negative bacteria, respectively. We showed that none of the tools outperform the others in all the studied features and that users should carefully choose one or another, or even combining them, depending on the particularities of their projects.

Network

Authors:

Gautam, J., Bowling Green State University

As a pioneer in making molecules in a sustainable way by strain engineering and industrial fermentation, Amyris has invested heavily in technology to develop strains and bring valuable molecules to the market at unprecedented speed. In the past three years, we have developed the Automated Scientist, a suite of mathematical models and machine learning algorithms that can iteratively perform the Design-Build-Test-Learn cycle for automated strain design and improvement. Thus far, the Automated Scientist has successfully created > 50,000 strains producing over 190 molecules. The design module of the Automated Scientist incorporates genome-scale metabolic models, an in-house knowledge store of design elements annotated with public data, and proprietary design algorithms to create pathway and strain designs. The pathway design to a target molecule can be visualized in the open-source tool Escher (first extended for industrial application at Amyris in 2014 by the author of the code) for route curation and additional flux balance simulation by biologists. To facilitate the exchange of knowledge between human scientists and the Automated Scientist, we have recently developed the Strain Optimization API. The API allows human scientists to specify the intended strain designs at a high level of abstraction, to validate their designs *in silico*, and to visualize the genotype-phenotype relationships. The Automated Scientist can parse the high-level input and automatically design the genetic architecture of the strains. In addition, the API can be accessed by Amyrisâ€™ high-throughput omics pipeline to identify strain-specific gene/protein/metabolite targets for testing up to 5,000 samples/day, and thus seamlessly accelerate the strain re-design process.

Developing a seamless pipeline from 13C labeling data to kinetic models with a genome-wide coverage

Authors:

Maranas, C. D., The Pennsylvania State University

Kinetic models of metabolic networks offer the potential of truly predictive models of metabolism. The mechanistic characterization of enzyme catalyzed reactions allows for accurate prediction of perturbations in metabolite concentrations and reaction fluxes in response to genetic and environmental perturbation that are beyond the scope of stoichiometric models. Despite their potential, the application of kinetic models to microbial strain design and metabolic engineering has been limited by (i)the paucity of fluxomic datasets that span the entire metabolism needed for parametrization (ii)the computational expense associated with kinetic model parameterization.

In this talk, we highlight a two-step pipeline that promises to alleviate some of the inherent computational challenges. It requires as an input a curated genome-scale-metabolic(GSM) model and 13C-labeling distributions for intracellular metabolites under multiple genetic and environmental perturbations. Additional data such as biomass yield, metabolite concentrations, enzyme kinetics and/or fermentation data can be integrated whenever available. The key concept here is to use the same GSM metabolic map for performing both metabolic flux elucidation and kinetic model parameterization. This eliminates any errors due to *ad-hoc*model aggregation/simplification. The first-step involves the application of improved algorithms for performing both stationary and instationary MFA using atom mapping model with genome-wide coverage. The second-step uses as input the elucidated metabolic flux ranges to estimate kinetic parameters that agree with all fluxomic datasets using the newly developed K-FIT-parameterization algorithm. The K-FIT-algorithm relies on a combination of nested decomposition schemes and iterative solution techniques to evaluate steady-state fluxes in response to genetic perturbations. It achieve orders of magnitude CPU improvements compared to a genetic-algorithm(GA)-based procedure. The approach has been tested for a core model of $E.coli$containing 109-reactions and 61-metabolites, a large-scale E.coli model with 319-reactions and 276-metabolites as well as models for clostridia. Using this workflow, 13C-data can be seamlessly translated into parameterized kinetic models.

Introducing thermodynamics into ME-models and crowders into kinetic models

Authors:

Hatzimanikatis, V., Swiss Federal Institute of Technology (EPFL)

The field of COBRA has impressively evolved since the first meeting in Iceland 7 years ago. The recent trends are moving towards the introduction of constraints that capture mRNA and protein expression (ME-models), thermodynamics and metabolomics (TFA), and most recently kinetics. While these additional constraints expand the capabilities and the scope of the models, they also introduce computational and modeling challenges. We will discuss here these challenges and we will present some new and computationally efficient problem formulations that allow the introduction of thermodynamic constraints into ME-models, and they also consider the effect of intracellular crowding conditions.

Machine Learning for Uncertainty Reduction in Biochemical Kinetic Models

Authors:

Miskovic, L., EPFL
Beal, J., EPFL
Moret, M., Harvard Medical School
Hatzimanikatis, V., Swiss Federal Institute of Technology (EPFL)

The primary goal of kinetic models is to capture the systemic properties of the metabolic networks, and we need large-scale kinetic models for reliable *in silico* analyses of the complex dynamic behavior of metabolism. However, parameter uncertainty hinders the development of kinetic models and uncertainty levels increase with the model size. Current methods for building kinetic models within constraint-based modeling frameworks address uncertainty indirectly by integrating data from different biological origins. We recently proposed iSCHRUNK, a computational approach which combines Monte Carlo sampling methods and machine learning techniques to characterize the uncertainties and to reveal complex relationships between the kinetic parameters and the responses of the metabolic networks. Monte Carlo sampling methods allow us to exploit synergies between different data sources and generate a population of kinetic models that are consistent with the available data and physicochemical laws. The machine learning allows us to data-mine the *a priori*generated kinetic parameters together with the integrated datasets and derive values of kinetic parameters consistent with the observed physiology. In this work, we modified iSCHRUNK to address a design question: can we identify the kinetic parameters and their values that give rise to a desired metabolic behavior? Such information is important for a wide variety of studies ranging from biotechnology to medicine. As an illustration, we used the proposed methodology to find parameters that ensure a rate improvement of the xylose uptake (XTR) in a glucose-xylose co-utilizing*S. cerevisiae* strain. Our results indicate that only three kinetic parameters need to be accurately characterized to reduce the uncertainty, and ultimately increase confidence in the design and control the metabolism desired responses. This framework paves the way for a new generation of methods that will systematically integrate the wealth of omics data and extract the information necessary for metabolic engineering and synthetic biology decisions.

Expanding Metabolic Models to Three Dimensions

Authors:

Brunk, E., University of California, San Diego
With the increasing coverage of genomic, proteomic and metabolomic data, we are witnessing the birth of a new era in biology- one which will change the face of medical research and clinical practice in unprecedented ways. A significant challenge that impedes further advancement of genomic medicine is understanding how to effectively integrate and translate big biological data into transformative health practices that impact patients lives. Recently, we have integrated protein structural information with the human metabolic network which provides a compelling new angle and scale for studying disease (Brunk et al. Nature Biotechnology 2018). Expanding reconstructions in this way provides new avenues for understanding how biochemical processes relate to mechanisms at the atomic scale. In this talk, I will discuss several recent applications that exemplify a new scientific frontier which has been termed "structural systems biology". For the first time, relationships between human metabolic genes, their encoded proteins, and reactions they catalyze can be described in the context of specific 3D configurations, interactions, and properties. Such capabilities open the door to exploring the spatial relationships of cancer mutations, the influence of SNPs on drug or metabolic responses, as well as the mechanistic underpinnings of basic biological processes like protein translation.

Key Decisions for the Development of the Next-Generation of Context-Specific Models

Authors:

Richelle, A., University of California, San Diego
Lewis, N., University of California, San Diego
Chiang, A., University of California, San Diego
Gutierrez, J., University of California, San Diego
Joshi, C., University of California, San Diego
Kellman, B., University of California, San Diego
Li, S., University of California, San Diego
Liu, J., University of California, San Diego

Genome-scale metabolic models provide a valuable context for analyzing data from diverse high-throughput experimental techniques. Since some enzymes are only active in specific environments, several algorithms have been developed to build context-specific models. However, the content and associated predictive capacity of resulting models is considerably impacted by the different assumptions used at each step of the extraction process: from the decisions on how to overlay data onto networks to the parameter choices for extraction algorithms[1,2]. These choices lead to a poor consensus in generated models, which may limit the use of context-specific methods for data-driven hypothesis.

We present a comparative analysis of existing extraction algorithms and an assessment of the key decisions influencing the data contextualization methods. From this work, we propose an approach to obtain better consensus across existing extraction algorithms and more accurate models. This approach is enabled with a framework we built for inferring metabolic functions that should be active in a specific context directly from transcriptomic data. These functions can be used to protect associated biochemical reactions during the implementation of extraction methods. The promising results obtained using this protectionist approach underline the potential interest of describing genome-scale metabolic reconstructions as more than a network of reactions but rather as an interconnected map of cellular functionalities.

1. Opdam, S., Richelle, A., Kellman, B., Li, S., Zielinski, D. C., & Lewis, N. E. (2017). A Systematic Evaluation of Methods for Tailoring Genome-Scale Metabolic Models. *Cell Systems*, *4*(3).http://doi.org/10.1016/j.cels.2017.01.010

2. Richelle, A., Joshi, C., & Lewis, N. E. (2018). Assessing key decisions for transcriptomic data integration in biochemical networks. *BioRxiv*, 301945.http://doi.org/10.1101/301945

Metabolic Coordination through Metabolite-Protein Interactions

Authors:

Sauer, U., ETH Zurich

How do bacteria know what goes on in their environment and how to they make appropriate decisions? While some bona fide extracellular sensors are known, there are far more environmental conditions and cellular responses than could possibly be dealt with through dedicated sensors. Instead, most microbial responses are based on intracellular changes to environmental changes. One of the first affected networks to just about any extracellular change is metabolism that passively responds to nutritional or chemical/physical challenges. Since fluxes and intracellular metabolite levels respond within seconds, allosteric binding of metabolites to regulatory proteins and enzymes is a highly effective and rapid sensing mechanism. Different from well-establish methods to assess physical interaction between proteins and between proteins and nucleic acids, however, methods to assess metabolite-protein interactions are still in its infancy. In this talk I will focus first on reconstructing the regulatory network of metabolite-enzyme interactions from databases, which currenty consists of 1500 unique regulatory interactions (1). I will then describe out efforts to experimentally map this network out further. The current results indicate that the known interactions are only the tip of the iceberg (2). Beyond knowing the interaction topology, I will conclude this talk on the even more challenging and conceptual problem: understanding which of the many regulation mechanisms actually matter for a given adaptation to elicit an appropriate physiological system response.

1. Reznik, Christodoulou, Goldford, Briars, Sauer, Segre & Noor. **Cell Reports** 20: 2666-2677 (2017).
2. Piazza, Kochanowski, Cappelletti, Fuhrer, Noor, Sauer & Picotti. **Cell**72:358-372 (2018).

An Upper Limit in Gibbs Energy Dissipation Governs Cellular Metabolism

Authors:

Heinemann, M., University of Groningen
Niebel, B., University of Groningen
Leupold, S., University of Groningen

The principles governing cellular metabolic operation are still poorly understood. Because very diverse organisms show relatively comparable physiologies, we hypothesized that a fundamental thermodynamic constraint might govern cellular metabolism. To investigate this, we developed a novel constraint-based model for*Saccharomyces cerevisiae* with a comprehensive description of the biochemical thermodynamics and including a Gibbs energy balance. Nonlinear regression analyses of metabolome and physiology data revealed the existence of an upper rate limit for the cellular Gibbs energy dissipation. Applying this limit in flux balance analyses using growth maximization as objective, the model correctly predicted physiologies, intracellular metabolic fluxes, as well as even some metabolite levels, for different glucose uptake rates. Our work indicates that cells arrange their metabolic fluxes such that with increasing substrate uptake rates, an optimal growth rate is accomplished, but the critical rate limit in Gibbs energy dissipation is never exceeded. Once all possibilities for further intracellular flux redistribution towards â€˜savingâ€™ Gibbs energy dissipation are exhausted, cells reach their maximal growth rate. We found that this principle also holds for *Escherichia coli* and on different carbon sources. Our work suggests that metabolic reaction stoichiometry, a limit in the cellular Gibbs energy dissipation rate, and the growth maximization objective might shape metabolism across different organisms and conditions.

Compass: FBA-Powered Modeling of Single-Cell RNA-Seq Characterizes Cell-to-Cell Metabolic Heterogeneity and Reveals Novel Therapeutic Targets in Autoimmunity

Authors:

Wagner, A., University of California, Berkeley
Detomaso, D., University of California, Berkeley
Wang, C., Brigham and Women's Hospital
Fessler, J., Harvard Medical School and Brigham and Women's Hospital,
Koul, A., Harvard Medical School and Brigham and Women's Hospital,
Yosef, N., University of California, Berkeley
Regev, A., Broad Institute of MIT and Harvard,
Kuchroo, V. K., Broad Institute of MIT and Harvard

The rapid advance in single-cell RNA-Sequencing (scRNA-Seq) is one of the most exciting recent developments in biomedical research. By comprehensively quantifying the transcriptome of individual cells, rather than the average over many cells, scRNA-Seq allows, for example, to discover rare cell sub-types and to retrace intermediate steps in lineage development. However, scRNA-Seq requires tailor-made computational methods that take advantage of its novel properties while accounting for its unique technical limitations (e.g., dropouts), and its immense data magnitude, which is fast approaching millions of cells per experiment

(Wagner et al., Nature Biotechnology 2016). Here, we answer these challenges in the realm of cellular metabolism. We present COMPASS, an FBA algorithm for comprehensive characterization of single-cell metabolic states. COMPASS capitalizes on the magnitude of scRNA-Seq data by modeling cells as points in a high-dimensional metabolic space, while mitigating scRNA-Seq data sparsity by smoothing kNN neighborhoods. The resulting metabolic space is directly interpretable in mechanistic terms. It allows data-driven characterization of metabolic heterogeneity among cells and associating metabolic programs with phenotypes of interest. To demonstrate COMPASS, we study T helper 17 (Th17) cells, which are remarkably plastic and mediate both autoimmunity and immune tolerance. A COMPASS characterization of Th17 metabolic heterogeneity recovered known associations of Th17 effector states with metabolic programs and predicted novel metabolic intervention targets. We show that the glycolytic shift in pathogenic Th17 is associated with extensive remodeling of anabolic pathways, which depends on pyruvate dehydrogenase (PDH). We then implicate the polyamine pathway, which has been scarcely studied in autoimmunity, in Th17 pathogenicity. *In vivo* deletion of either PDH or polyamine synthesis led to considerably better clinical outcome in a murine model of multiple sclerosis (MS). The clinical significance of our findings is supported by differential abundance of polyamines in the blood of healthy human donors compared with MS patients.

Metabolic Model-Based Evaluation of Microbiome-Metabolome Association Studies

Authors:

Noecker, C., University of Washington
McNally, C., University of Washington
Chiu, H. C., MediaTek
Borenstein, E., University of Washington

Identifying specific microbial drivers of variation in metabolic phenotypes is a major goal in the study of host-associated and environmental microbiomes. Correlation-based analysis of paired microbiome-metabolome datasets is a widespread approach to this objective. To date, however, the efficacy and limitations of this approach have not been evaluated. To address this challenge, we developed a mathematical definition of the contribution of each taxon to metabolite variation based on its uptake and secretion fluxes. We applied a multi-species dynamic genome-scale metabolic modeling pipeline to simulate simplified gut communities, generating idealized microbiome-metabolome datasets with precisely known microbial metabolic fluxes. Comparing the observed taxon-metabolite correlations in this simulated setting with the true taxonomic contributors, we found that correlation-based analysis poorly predicts key contributors, with low accuracy and a high false discovery rate. Importantly, however, the predictive value was strongly influenced by both metabolite and taxon properties, as well as exogenous environmental variation. These findings have practical implications for the analysis and interpretation of microbiome-metabolome studies. This study illustrates the utility of multi-scale metabolic modeling as a simulation tool to inform study design and methods development in microbiome research.

Yeast-GEM: Reviving the Consensus Genome-Scale Model of S. Cerevisiae As a Standard in the Community

Authors:

Sánchez, B. J., Chalmers University of Technology
Lu, H., Chalmers University of Technology
Li, F., Chalmers University of Technology
Kerkhoven, E. J., Chalmers University of Technology
Nielsen, J., Chalmers University of Technology
Domenzain, I., Chalmers University of Technology

Genome-scale models (GEMs) are essential tools for understanding and computing metabolism. However, as they are so large (thousands of metabolites and reactions), it becomes challenging to keep track of changes between two versions of the same model. Furthermore, a framework is needed for researchers to work on a single model at the same time, without creating conflicts.

Here, we have brought back to life the GEM for *Saccharomyces cerevisiae* as yeast-GEM: a version controlled model publicly available for open collaboration. Changes are tracked with Git, can be easily visualized on Github, and are linked to complementary data and scripts, so anyone can know what was changed, who did it and why. By hosting the model in Github we also allow model developers to work on parallel, we periodically release new versions of the model, and we publicly display current and future work, organized in different projects. Additionally, we take advantage of memote, the metabolic model test suite, to display the quality of our model as we work on it.

We will present numerous model improvements that we have performed using this framework, such as an extensive curation to metabolites, reactions and genes based on new genome annotation; the introduction of SLIME reactions, which Split Lipids Into Measurable Entities for properly representing limitations on both the lipid classes and at the same time the acyl chains; and a growth validation study under different carbon/nitrogen/sulphur/phosphate sources.

Finally, we will show how we connect this model to GECKO, a toolbox for adding enzyme constraints to a GEM, and achieve a continuous synchrony of the metabolic model (yeast-GEM) and the enzyme-constrained model (ecYeast-GEM), to offer to the community a set of ready to use constrained-based models of yeast. See more of yeast-GEM, and contribute if you wish, athttps://github.com/SysBioChalmers/yeast-GEM

Delineating Metabolic Reprogramming and Redox Homeostasis in Antibiotic Resistant Chromobacterium Violaceum Using Constraint-Based Analysis

Authors:

Raghunathan, A., CSIR - National Chemical Laboratory
Banerjee, D., CSIR - National Chemical Laboratory

Metabolic network reconstruction is central to many aspects of metabolic analysis and engineering, and is essential for ongoing efforts towards mechanistic microbial ecosystem modeling. In general, the reconstruction process involves translating genomic information into a metabolic network that can provide insight into cell-wide metabolic fluxes and other phenotypes. A number of well-curated reconstructions exist for different model organisms, but a major challenge in the field is to reconstruct metabolic networks for diverse environmentally and medically relevant organisms. Various data sources including genomic, metabolomic, and phenotypic data can be used to guide metabolic network reconstruction. Additionally, there are several methods available for automatically generating draft reconstructions, including by using probabilistic genome annotations. However, few methods have directly attempted a comprehensive, automated integration of multiple data types, while at the same time explicitly dealing with the inherent uncertainty in the reconstruction process. We have developed a Markov chain Monte Carlo method to sample metabolic networks from a probability space shaped by the available data. Our method formalizes the data integration process and clarifies the representation of uncertainty by generating a natural ensemble of metabolic network reconstructions. As a proof of concept, we implemented this method on simulated genomic and metabolomic data from the *E. coli* iJO1366 metabolic network. Moving forward, we are working on including additional data types, such as phenotypic assays, with the goal of applying this method to a large group of environmentally-relevant marine microbes.

Urea cycle dysregulation, emerging pyrimidines mutation bias and enhanced response to immunotherapy in cancer

Authors:

Lee, J. S., Cancer Data Science Lab, NCI/NIH
Ruppin, E., University of Maryland

The urea cycle (UC) is the main metabolic pathway by which mammals dispose waste nitrogen. Here we show that the expression of most UC enzymes is altered in many tumors, leading to a general metabolic hallmark that we term UC dysregulation (UCD). UCD elicits nitrogen diversion towards carbamoyl-phosphate synthetase2, aspartate transcarbamylase and dihydrooratase (CAD) activation and enhances pyrimidine synthesis, resulting in detectable changes in nitrogen metabolites in both patient tumors and bio-fluids. The accompanying excess of pyrimidine vs purine nucleotides results in a novel genomic signature consisting of transversion mutations at the DNA, RNA and protein levels. This mutational bias is associated with more hydrophobic tumor antigens and with a better response to immune checkpoint inhibitors, independent of mutational load. Taken together, our findings demonstrate that UCD is a common feature of tumors, which profoundly impacts carcinogenesis, mutagenesis and immunotherapy response.

[Joint work with the Erez lab at the Weizmann Institute, Israel]

Modeling Physiological Responses to Stress: Metabolism of Fasting and Acetaminophen-Induced Liver Damage in the Laboratory Rat

Authors:

Wallqvist, A., U.S. Army Medical Research and Materiel Command
Pannala, V., U.S. Army Medical Research and Materiel Command
Wall, M., Vanderbilt University School of Engineering
Estes, S., Department of Molecular Physiology and Biophysics
Irina, T., Vanderbilt University School of Engineering
Tracy, O., Vanderbilt University School of Engineering
Richard, P., Department of Molecular Physiology and Biophysics
Kalyan, V., U.S. Army Medical Research and Materiel Command
Reifman, J., U.S. Army Medical Research and Materiel Command
Shiota, M., Vanderbilt University School of Medicine
Young, J., Vanderbilt University School of Engineering

Metabolism is an integral component of our response to physiological stress. We need to convert nutrients to energy, make the requisite macromolecular components, and use these components, to function normally and handle varying physiological stress. Capturing these physiological responses using metabolic network modeling is instrumental for obtaining scientific insights from transcriptomic and metabolomic data. Here, we studied coupled liver and kidney metabolism in the laboratory rat under mild stress conditions, i.e., during short-term fasting and during the initial period of acute acetaminophen toxicity before the onset of cellular damage. We used a genome-scale network reconstruction of rat liver and kidney metabolism, together with a modeling approach constrained by stress-induced gene expression data, central carbon fluxes derived from isotope tracer techniques, and physiological flux bounds. We gauged the performance of the approach by comparing metabolite levels secreted in plasma and urine predicted by the model with those obtained from non-targeted metabolic profiling. In the case of short-term fasting, we correctly accounted for glucose/glycogen metabolism in the liver, using a modeling approach based on the metabolic-network structure coupled only with measured central carbon fluxes. For acetaminophenâ€"a known hepatotoxicantâ€"we further incorporated changes in liver and kidney gene expression into multiple modeling frameworks to study the resultant changes in endogenous metabolism. A comparison of the modeled results with the global metabolic profiling data revealed that, our approach satisfactorily predicted altered plasma metabolite levels as early as 5 h after exposure to 2 g/kg of acetaminophen, and that, after 10 h of exposure the predictions significantly improved. We achieved a 70% correspondence between predicted and measured changes of metabolite levels in plasma and urine. The coupled multi-tissue modeling framework of *in vivo* metabolism provides both mechanistic insights and a capability to identify plasma and urine metabolites as early markers of toxicant-induced organ damage.

Differential Metabolic Functionality in Antibiotic-Resistant Pseudomonas aeruginosa Revealed with an Integrated Computational and Experimental Approach

Authors:

Dunphy, L. J., University of Virginia
Papin, J., University of Virginia
Yen, P., University of Virginia

Changes in bacterial metabolism accompanying the development of antibiotic resistance remain poorly understood. In this study, we performed a single-carbon source utilization screen on lab-evolved antibiotic-resistant *Pseudomonas aeruginosa* to investigate these changes. The metabolic capabilities of piperacillin-resistant, tobramycin-resistant, and ciprofloxacin-resistant *P. aeruginosa* as well as paired ancestral and media-evolved control lineages were evaluated by measuring growth curves on 190 unique carbon sources. Our resulting 950 growth curves revealed that resistant lineages exhibit mainly decreased catabolic function with occasional gains of function compared to antibiotic-sensitive *P. aeruginosa*. In resistant lineages we also observed changes in growth dynamics, including growth rate and time to reach mid-exponential phase. A genome-scale metabolic network reconstruction of *P. aeruginosa* strain UCBPP-PA14, iPau1129, was used to contextualize whole-genome sequencing data of the resistant lineages. The model was used to predict the impact of resistance mutations on loss of catabolic function. For example, five genes deleted in the piperacillin-resistant lineage were predicted to drive loss of the ability to utilize L-leucine. Model predictions were experimentally validated with a transposon mutant library. Our results show that metabolism is altered through the evolution of antibiotic resistance. Instances where our model failed to correctly predict genotype-phenotype relationships highlight gaps in our current understanding of *P. aeruginosa*metabolism. Our combined computational and experimental framework can be applied to identify metabolic limitations in other antibiotic-resistant pathogens. Drug-driven metabolic limitations have the potential to be targeted to select against antibiotic-resistant populations.

Model-Based Prediction of Functional SNPs Suggests Factors for Metabolic Diversity and Drug Resistance across Human-Associated Mycobacterium Tuberculosis

Authors:

Øyås, O., ETH Zurich
Stelling, J., ETH Zuerich
Sauer, U., ETH Zurich
Trauner, A., Swiss TPH
Borrell, S., University of Basel
Zampieri, M., ETH Zurich
Gagneux, S., Swiss TPH
Zimmermann, M., ETH Zurich

The *Mycobacterium tuberculosis* complex (MTBC) is a group of closely related pathogenic bacteria that can cause tuberculosis (TB) upon infection. TB is a leading cause of human mortality worldwide and threatens to remain so for the foreseeable future due to the emergence of multidrug-resistant MTBC strains. A characteristic feature of the MTBC is that it harbors comparatively little genetic variation, but a large share of its single-nucleotide polymorphisms (SNPs) are non-synonymous and experimental evidence is challenging the long-held assumption that phenotypic diversity in the MTBC is negligible. Bridging the genotype-phenotype gap in the MTBC could therefore provide valuable insights into the evolution of phenotypes such as drug resistance.

To address this problem, we built a single constraint-based model that integrates the exometabolomes and genomes of 18 MTBC strains from six lineages native to different parts of the world with strain-specific genome-scale metabolic models. We used the model to predict the metabolic effects of non-synonymous SNPs in enzyme-encoding genes via a three-step optimization approach, aiming to explain as much of the observed metabolic variation as possible by as few SNPs as possible while obtaining consistent flux distributions. Using our predicted SNP effects, we classified 88 SNPs (15%) as functional. These functional SNPs affect 67 unique enzymes across most metabolic pathways and include a SNP in pyruvate kinase previously shown to be functional in *Mycobacterium bovis*. In addition, we predicted three functional SNPs in enzymes involved in folate metabolism and we suggest a possible explanation for differential sensitivity to *para*-aminosalicylic acid, one of the antibiotics currently used to treat multidrug-resistant TB. Concluding, our method is capable of predicting the metabolic effects of genetic variation in microbes and allowed us to connect genetic and metabolic diversity in the MTBC.

Synthetic Lethality in Cancer Research Via Genetic Minimal Cut Sets

Authors:

Planes, F. J., TECNUN, University of Navarra
Apaolaza, I., TECNUN, University of Navarra
Valcárcel, L. V., CIMA, University of Navarra
Agirre, X., CIMA, University of Navarra
Prosper, F., CIMA, University of Navarra
San Jose, E., CIMA, University of Navarra

Synthetic lethality is a promising approach in precision medicine and cancer as it largely expands the number of possible drug targets and creates an opportunity for selectivity. The increasing evidence of metabolic reprogramming of cancer cells makes it ideal to exploit the concept of synthetic lethality. A number of in-silico tools have been developed to target cancer metabolism using a synthetic lethality approach. In particular, constraint-based modeling (CBM) for genome-scale metabolic networks has received much attention. Here, we present a novel CBM approach to synthetic lethality that is based on the concept of genetic Minimal Cut Sets (gMCSs). With respect to existing methods, our approach avoids the step of network contextualization and integrates â€“omics data in a more natural and objective manner. In addition, it enables not only the detection metabolic targets but also response biomarkers for them. To illustrate our approach, we first show the results of an experimental proof-of-concept in multiple myeloma (MM), where we validated the therapeutic potential of RRM1 inhibition in different MM cell lines. We also predicted a metabolic signature based on gene expression data that explained the response to RRM1 inhibition in different cancer cell lines. Second, we show preliminary results of a study where response biomarkers for the effectiveness of Methotrexate in different cancer types is predicted following our methodology. This new algorithm, freely available in the COBRA Toolbox, undoubtedly opens new avenues to develop precision medicine strategies in complex and unaddressed clinical questions involving heterogeneous molecular data.

Identifying and Targeting Key Cellular Mechanisms for Proliferation in Malaria Parasites: A Combined Experimental and Computational Strategy

Authors:

Chiappino-Pepe, A., Swiss Federal Institute of Technology (EPFL)
Bushell, E., Wellcome Trust Sanger Institute
Limenitakis, R., University of Bern
Pandey, V., Swiss Federal Institute of Technology (EPFL)
Rayner, J. C., Wellcome Trust Sanger Institute
Heussler, V., University of Bern
Billker, O., Wellcome Trust Sanger Institute
Hatzimanikatis, V., Swiss Federal Institute of Technology (EPFL)

Recent advances in cell genome editing techniques enable the generation of high-throughput gene knockout data in the malaria parasites *in vivo*. Integrative analysis of this data can lead to the identification of biological mechanisms that explain the observed phenotypes and that provide testable hypotheses for further discoveries. Metabolic modeling can cope with the tangled metabolism of the malaria parasites, and hence is a compelling approach for understanding the parasites physiology.
In this study, we present a combined experimental and computational approach that suggests cellular mechanisms for targeting the malaria parasites. We predict *in silico* and test*in vivo* lethal knockouts and synthetic lethal pairs in the blood and liver stages of the malaria infection.

We perform computational analyses on a newly developed genome-scale model of the malaria parasite *Plasmodium berghei* (iPbe), and we use high-throughput gene knockout data generated in the PlasmoGEM project. The comparison between data and gene essentiality predictions allow the understanding of the parasiteâ€™s physiology in the blood and liver stages. We identify the thermodynamic bottlenecks, genetic interactions, and the accessibility to nutrients behind the phenotypes. When we simulate in iPbe the hypothesized physiology, we achieve 80% consistency between the prediction of essential genes and the experimental data. This result indicates that our model iPbe is a valuable framework for the generation of testable hypothesis on antimalarial targets. Overall, the knowledge generated in this framework will serve to tackle more efficiently the malaria parasitesâ€™ metabolism during infection.

Inferring Metabolic Mechanisms of Interaction within a Defined Gut Microbiota

Authors:

Medlock, G. L., University of Virginia
Carey, M. A., University of Virginia
Kolling, G., University of Virginia
Swann, J., Imperial College London
Giallourou, N., Imperial College
Mundy, M., Mayo Clinic
McDuffie, D., University of Virginia
Papin, J. A., University of Virginia

Within the gastrointestinal tract of mammals, the diversity and number of species present enable a multitude of metabolic interactions. Methods for predicting these interactions are attractive because they may enable engineering of microbiome function. However, identifying the mechanism and consequences of metabolic interactions between even two species is incredibly challenging. In this work, we developed, applied, and experimentally tested a framework for identifying potential metabolic mechanisms associated with interspecies interactions.

We performed pairwise co-culture growth experiments using bacterial species a model mouse microbiota. After 72 hours of growth, we quantified the abundance of each strain and determined metabolite consumption and production using untargeted supernatant metabolomics. We applied our framework, which we call the Constant Yield Expectation

(ConYE) model, to dissect emergent metabolic behaviors that occur in co-culture.

The ConYE model assumes the yield of a metabolite is constant in monoculture and co-culture. Using ConYE, we identified widespread indications of increased efficiency of biomass production in co-culture. We interrogated an amino acid crossfeeding interaction that is likely to confer a growth benefit to one ASF strain (Clostridium sp. ASF356) in coculture with another strain (Parabacteroides goldsteinii ASF519). We attempted *in silico* identification of growth-enhancing metabolites by constructing genome-scale metabolic network reconstructions for the ASF strains. Using the simulations together with the ConYE results, we designed media supplementation experiments and verified that the proposed interaction leads to a growth benefit for this strain.
Our results reveal the types and extent of emergent metabolic behavior in microbial communities and demonstrate how metabolomic data can be used to identify potential metabolic. Although we focus on growth-modulating interactions, the framework we develop can be applied to generate specific hypotheses about mechanisms of interspecies interaction involved in any phenotype of interest within a microbial community.

Understanding the Role of Bile Acids in Alzheimer?s Disease

Authors:

Baloni, P., Institute for Systems Biology
Price, N. D., Institute for Systems Biology
Funk, C., Institute for Systems Biology
Simeonidis, E., Institute for Systems Biology
Yan, J., Indiana University School of Medicine
Nho, K., Indiana University School of Medicine
Arnold, M., Helmholtz Zentrum München, German Research Center for Environmental Health
Kastenmueller, G., Institute of Bioinformatics and Systems Biology
Louie, G., Duke Institute for Brain Sciences
Kueider-Paisley, A., Duke Institute for Brain Sciences
Saykin, A. J., Indiana University School of Medicine
Kaddurah-Daouk, R., Duke Institute for Brain Sciences

Alzheimerâ€™s disease (AD) is the leading cause of dementia. AD-related neurodegeneration has been correlated with metabolic dysfunction. An integrative analysis of transcriptomic, proteomic and metabolomic data from the brains of post-mortem AD patients and controls has provided evidence of genes and metabolites involved in altered metabolic pathways in AD. We have built context-specific draft metabolic networks for multiple brain regions, using high throughput transcriptome data of post-mortem brain samples from AD patients and controls, and integrated them with Recon3D, the most comprehensive and recent human metabolic reconstruction. We previously identified a role for circulating bile acids in AD, as well as upstream changes in cholesterol metabolism. Primary bile acids are synthesized in the liver, whereas secondary bile acids are typically produced by bacteria in the gut. Increased levels of secondary bile acids and ratios to their primary bile acid educts have been linked to AD and cognitive decline. Expression analysis shows that genes involved in alternative bile acid synthesis pathways are expressed in the brain, while genes in the classical pathway are not, thus providing insights into bile acid metabolism in the brain. By integrating transcriptomics data and metabolomics measurements of bile acids from brain tissue of AD patients and controls within these metabolic networks, our models can capture *in silico* changes in these pathways, highlighting the role of brain bile acid metabolism in AD pathophysiology. Our brain-tissue metabolic models can be exploited to capture *in silico* changes and possibly identify metabolic biomarkers prior to disease manifestation, making them useful for identifying at-risk individuals. Using the information of brain-specific transcriptional regulatory networks, we aim to identify key genes involved in regulating metabolic changes. Our approach combines multi-omics information and provides insights into interactions of primary and secondary bile acids and their possible connection to cognitive decline and AD.

The Modular Tumor Checkpoint Landscape of Human Cancer

Authors:

Paull, E., Columbia University
Califano, A., Columbia University

Background: Since the discovery, made in 1905 by Schaudinn and Hoffmann, of Treponema pallidum ssp. pallidum as the etiologic agent of syphilis, medicine has made significant progress against this disease. Yet, despite the availability of diagnostic tests and a therapy based on antibiotics, the world has not stopped being burdened by syphilis, that has been re-emerging globally over the last few decades (WHO Report, 2008), for which no vaccine is still available and which, moreover, in its early stages enhances the transmission of HIV. Continuous in vitro culture of this organism has still not been achieved, imposing a substantial roadblock to its experimental inspection, and even the sequencing of its genome (Fraser et al., 1998) did not yield an obvious solution to the cultivation problem. While much has already been tried on the laboratory bench, this pathogen has still not (to our knowledge) been tackled using a systems biology approach.

Results: Here, we present a first manually curated draft reconstruction of the metabolic network in Treponema pallidum ssp. pallidum towards a genome-scale metabolic model (GEM). At this time, the model iSM161 comprises 161 genes of 1,039 predicted open reading frames that are responsible for 239 reactions of 277 metabolites. The model is still under development and steadily updated. For the reconstruction, COBRApy has been used, where subsystem information is added and parsed as SBML groups extension using libSBML.

Discussion: Using this reconstruction, together with the application of COBRA methods, we anticipate to gather new insights into the pathogenâ€™s physiology and pathology, and in how this spirochete, which has earned the designation of â€œstealth pathogen,â€ succeeds in making a living and eluding humanâ€™s immune defenses as well as cultivation attempts. It is planned to make the model available to the community in SBML format.

J

As a pioneer in making molecules in a sustainable way by strain engineering and industrial fermentation, Amyris has invested heavily in technology to develop strains and bring valuable molecules to the market at unprecedented speed. In the past three years, we have developed the Automated Scientist, a suite of mathematical models and machine learning algorithms that can iteratively perform the Design-Build-Test-Learn cycle for automated strain design and improvement. Thus far, the Automated Scientist has successfully created > 50,000 strains producing over 190 molecules. The design module of the Automated Scientist incorporates genome-scale metabolic models, an in-house knowledge store of design elements annotated with public data, and proprietary design algorithms to create pathway and strain designs. The pathway design to a target molecule can be visualized in the open-source tool Escher (first extended for industrial application at Amyris in 2014 by the author of the code) for route curation and additional flux balance simulation by biologists. To facilitate the exchange of knowledge between human scientists and the Automated Scientist, we have recently developed the Strain Optimization API. The API allows human scientists to specify the intended strain designs at a high level of abstraction, to validate their designs *in silico*, and to visualize the genotype-phenotype relationships. The Automated Scientist can parse the high-level input and automatically design the genetic architecture of the strains. In addition, the API can be accessed by Amyrisâ€™ high-throughput omics pipeline to identify strain-specific gene/protein/metabolite targets for testing up to 5,000 samples/day, and thus seamlessly accelerate the strain re-design process.

Towards a more complete view of a cell's functions through genome scale models of metabolism, protein synthesis, post-translational modification, and secretion

Authors:

Lewis, N., University of California

In mammalian cells, metabolism is a core process driving homeostasis, but variations in other cell processes largely define cell type identity and cell-type specific functions. The profile of secreted and membrane proteins show substantial cell-type specificity and drive many tissue specific functions. These proteins, encoded by up to 1/3 of mammalian protein-coding genes, include hormones, membrane proteins, and enzymes that modulate the extracellular space, and these are synthesized and trafficked through the secretory pathway. The pathway complexity, however, obfuscates its impact on the secretion of different proteins. Unraveling its impact on diverse proteins is particularly important since the pathway is implicated in many diseases and harnessed for biopharmaceutical production. We have delineated the core secretory pathway functions and integrated them with genome-scale metabolic models of human, mouse, and Chinese hamster ovary cells. Here I will discuss our efforts to use this conceptual expansion of constraint-based metabolic modeling to obtain insights into bioenergetic demands imposed by protein synthesis and secretion, and how these models can be deployed in efforts to engineer mammalian cells for enhanced secretion of high-value biologic drugs. Thus, this work represents a knowledge-base of the mammalian secretory pathway that serves as a novel tool for systems biotechnology.

Deciphering the Mechanisms Underlying Nutrient Exchange in Insect-Microbe Symbioses

Authors:

Ankrah, N., Cornell University
Wu, C., Cornell University
Douglas, A., Cornell University

Many insects of medical and agricultural importance depend on symbiotic associations with nutrient producing bacteria. Very commonly, net nutrient production by the bacteria is dictated by metabolic competition and cooperation among multiple bacterial taxa and strongly influenced by the metabolites derived from the insect host, but the mechanistic details of the multi-way metabolic interactions are poorly understood. To investigate the processes shaping insect-microbe metabolic interactions and to derive estimates of the composition and amount of nutrients exchanged between insect and microbe, we reconstructed genome and transcriptome-informed metabolic models of sap-feeding insects that harbor intracellular bacterial partners and the bacterial community in the Drosophila gut. Our simulations reveal that among intracellular bacterial partners the insect host controls bacterial production of nutrients, by precise controls over the concentrations of substrate metabolites in essential nutrient biosynthetic pathways. The net result is the structuring of the metabolic networks of the bacteria, to promote cooperative cross-feeding of metabolites and minimize competition for host-derived substrates. Among gut microbes competitive interactions dominate and the quality and quantity of nutrients available to the host is influenced by the composition of gut microbiota which synthesize and consume host and microbe derived nutrients. Our studies elucidate the metabolic basis of insect-microbe interactions and provide the methodology to identify specific gene targets for novel pest control strategies and to select the optimal microbial consortia for personalized microbial therapies in treating animal diseases associated with microbial dysbiosis.

Incorporation of Dynamic pH into an Integrative Model of a Nitrification Microcosm Co-Culture of Nitrosomonas Europaea and Nitrobacter Winogradskyi

Authors:

Chaplen, F., Oregon State University
Mellbye, B., Oregon State University

Understanding coupled biogeochemical systems is important for elucidating anthropogenic environmental impacts. Nitrifying bacteria are fundamental elements of the nitrogen (N) Cycle. N gases production by soils through microbial action can be a significant source of atmospheric NO_x and N_2O. Multi-level genome-scale modeling of*N. europaea* and *N. winogradskyi* in a microcosm suggests a key role for NO in the abiotic and biotic chemistry of nitifying systems[1]. Further refinement of the integrative model to incorporate dynamic pH profiling resulted in additional insights into the physiology of *N. europaea*. Matching the dynamic pH profile of the microcosm required the use of the Euclidean norm for flux distribution minimization during optimization of the genome-scale model component of the integrative model. The requirement for the Euclidean norm suggested that *N. europaea* in the microcosm operates in a manner that optimizes metabolite channeling[2]. Modeling with the enhanced integrative model also suggested that *N. europaea* facilitates conversion of ammonium ion to ammonia. A possible mechanism for ammonium conversion is uptake by ammonium transporter (*amtB*; NE0448) or active membrane-potential-driven mechanisms into a more basic cytoplasm[3].

[1] **Mellbye, B.L., Giguere, A., Murthy, G.S., Bottomley, P.J. Sayavedra-Soto, L.A. and Chaplen F.** (2018) Genome-scale, constraints-based modeling of nitrogen oxide fluxes during co-culture of *Nitrosomonas europaea* and*Nitrobacter winogradskyi*. mSystems 3(3): e00170-17. (Editorâ€™s Pick) Retrieved fromhttp://dx.doi.org/10.1128/mSystems.00170-17
[2] **Wu, H.-Q., Cheng, M.-L., Lai, J.-M., Wu, H.-H., Chen, M.-C., Liu, W.-H., Wu, W.-H., Chang, P., M.-H., Huang, C.-Y. H., Tsou, A.-P., Shaio, M.-S., and Wang, F.-S.** (2017) Flux balance analysis predicts Warburg-like effects of mouse hepatocyte deficient in miR-122a.*PLOS Comput. Biol.* **13**(7):e1005618. Https://doi.org/10.1371/journal.pcbi.1005618.
[3] **Cherif-Zahar, B., Durand, A., Schmidt, I., Hamdaoui, N., Matic, I., Merrick, M, and Matassi, G.**(2007) Evolution and functional characterization of the *RH50* gene from ammonia-oxidizing bacterium*Nitrosomonas europaea. J. Bacteriol.***189**(24):9090-2100. Doi:10.1128/JB.01089-07

Dynamic Side of the Warburg Effect: Glycolytic Intermediate Storage in Tumor Cells to Buffer Fluctuating Glucose and O2 Supply

Authors:

van Beek, J. H. G. M., VU University medical centre

Tumor cells show the Warburg effect: high glucose uptake and lactate production despite sufficient oxygen supply. A multi-scale computational model of tumor cell metabolism, intracellular metabolite storage and transport in tissue by blood flow and diffusion is developed here. This model is used to investigate the dynamic behavior of the metabolic system underlying the Warburg glycolytic phenotype. Remarkably, experiments have shown that the ascites tumor cells studied by Otto Warburg can transiently take up glucose an order of magnitude faster than the already high glycolytic rate that he measured for hours. This transiently very high glucose uptake is investigated here with a computational model that reproduces kinetic experiments on these ascites tumor cells. Model analysis of the data shows that the head section of the glycolytic chain in the tumor cells is partially inhibited in about a minute when substantial amounts of glucose have been taken up intracellularly. This head section of the glycolytic chain is subsequently disinhibited slowly when concentrations of glycolytic intermediates fall to low levels again. Based on these dynamic characteristics, simulations of tissue with fluctuating O_2 and glucose supply predict that tumor cells greedily take up glucose when this periodically becomes available at low concentrations, leaving very little for other cells. The glucose is stored as fructose 1,6-bisphosphate and other glycolytic intermediates, which are used for ATP production during O_2 and glucose shortages that last several minutes. This dynamic energy buffer capacity may have selective advantages for tumor cells experiencing cycling hypoxia and nutrient shortages, which are often found in cancer tissue. The hypothesis is put forward here that dynamic regulation of the powerful glycolytic enzyme system in tumor cells is used to buffer oxygen and nutrient fluctuations in tissue. The stored glycolytic intermediates can support the ATP requirements of the tumor cells for 2-10 minutes.

Toward a Proteome-Complete Model of the Human Red Blood Cell

Authors:

Yurkovich, J. T., University of California, San Diego
Yang, L., University of California, San Diego
Palsson, B. O., University of California, San Diego

The human red blood cell has served as a starting point for the application and development of systems biology approaches due to its simplicity, intrinsic experimental accessibility, and importance in human health applications. Here, we present a multi-scale computational model of the human red blood cell that accounts for metabolism and macromolecules. Proteomics data are used to place quantitative constraints on individual protein complexes that catalyze metabolic reactions, as well as a total proteome capacity constraint. We explicitly describe molecular mechanismsâ€"such as hemoglobin binding and the formation and detoxification of reactive oxygen speciesâ€"and account for sequence variations between individuals, allowing for personalized physiological predictions. This model allows for direct computation of the oxyhemoglobin curve as a function of model species and a more accurate computation of the flux state of the metabolic network. More broadly, this work represents another step toward building increasingly more complete systems biology whole-cell models.

Koptic: A Novel Approach for in silico Prediction of Enzyme Kinetics and Regulation

Authors:

Schroeder, W., University of Nebraska - Lincoln
Saha, R., University of Nebraska-Lincoln

Computational modeling of metabolism enables the design of engineering interventions directed to the overproduction of a specific bioproduct or improvement of plant performance. Flux Balance Analysis (FBA) is the primary tool used for this purpose, but has significant limitations due to the lack of reaction kinetics, chemical species concentration, and metabolic regulation. In contrast, kinetic models of metabolism (kMMs) provide not only a more accurate method for designing novel biological systems but also for the characterization of reaction kinetics, metabolite concentration, and metabolic regulation in these systems; however, the multi-omics data required for their construction is prohibitive to their development and widespread use.

Here, we introduce **K**inetic**OPT**imization using **I**nteger**C**onditions (**KOPTIC**), which can circumvent the omics data requirement and semi-automate kMM construction by using reaction rates and concentration data derived from a metabolic network model to return plausible kinetic mechanisms through an optimization-based approach.*Arabidopsis thalianaâ€™s* (hereafter*Arabidopsisâ€™s*) prominent role in â€˜omicsâ€™ and plant science research makes it an ideal organism for the verification of KOPTIC-predicted kinetic mechanisms. While, several metabolic network models for*A. thaliana* already exist, the core carbon metabolism of this organism was chosen as the test system. A four-tissues (leaf, root, seed, and stem) metabolic network model, was reconstructed for *Arabidopsis* (1015 reactions, 901 metabolites, 508 genes). FBA was performed at 71 time-points to simulate the *Arabidopsis*lifecycle, and KOPTIC was applied to the FBA data. In total KOPTIC predicted 3577 regulatory interactions with a median fit error of 13.44%. More than 30 verified by existing literature. This research showcases how an optimization-based approach can be used to create meaningful hypotheses of reaction kinetics and increase mechanistic understanding of metabolism.

New Directions for Cobra

Authors:

Palsson, B. O., University of California, San Diego
Calls for effective multi-omic data integration are growing, and with it, the need for explanatory-AI (or xAI) has arisen. COBRA methods provide a structured and fundamental basis for data analysis and interpretation and should offer opportunities for xAI broadly for the life sciences. Three new directions for COBRA will be discussed: 1. BD2K; 2. Structural proteomics as an additional omics data type; and 3. The pangenome goldmine.

Integrated Constraint-Based Analysis of an Engineered Violacein Pathway in Escherichia coli

Authors:

Banerjee, D., CSIR - National Chemical Laboratory
Immanuel, S. R. C., CSIR - National Chemical Laboratory
Rajankar, M. P., CSIR - National Chemical Laboratory
Raghunathan, A., CSIR - National Chemical Laboratory

Metabolic network reconstruction is central to many aspects of metabolic analysis and engineering, and is essential for ongoing efforts towards mechanistic microbial ecosystem modeling. In general, the reconstruction process involves translating genomic information into a metabolic network that can provide insight into cell-wide metabolic fluxes and other phenotypes. A number of well-curated reconstructions exist for different model organisms, but a major challenge in the field is to reconstruct metabolic networks for diverse environmentally and medically relevant organisms. Various data sources including genomic, metabolomic, and phenotypic data can be used to guide metabolic network reconstruction. Additionally, there are several methods available for automatically generating draft reconstructions, including by using probabilistic genome annotations. However, few methods have directly attempted a comprehensive, automated integration of multiple data types, while at the same time explicitly dealing with the inherent uncertainty in the reconstruction process. We have developed a Markov chain Monte Carlo method to sample metabolic networks from a probability space shaped by the available data. Our method formalizes the data integration process and clarifies the representation of uncertainty by generating a natural ensemble of metabolic network reconstructions. As a proof of concept, we implemented this method on simulated genomic and metabolomic data from the *E. coli* iJO1366 metabolic network. Moving forward, we are working on including additional data types, such as phenotypic assays, with the goal of applying this method to a large group of environmentally-relevant marine microbes.

The Metabolic Basis of Carbon Accumulation in Chlamydomonas Reinhardtii during Nitrogen Deprivation

Authors:

Clark, T., Michigan State University
Shachar-Hill, Y., Michigan State University
Juergens, M., Monsanto Company

The heterogeneity of the aromatic products from lignin catalytic depolymerization is one of the major challenges for lignin valorization. Microbes have evolved many catabolic pathways to tackle the challenge by funneling such heterogeneous intermediates to a few central aromatic products, which lead further to intra- or extradiol ring opening to produce value-added chemicals. However, such funneling pathways are only well-characterized in a few organisms such as*Sphingobium* sp. SYK-6 and *Pseudomonas putida* KT2440. Further effort to explore novel bacterial funneling pathways is necessary. Herein, we apply the *de novo*pathway design tool (novoStoic [1]) to computationally prospect all possible pathways for lignin-derived mono- and biaryls. First, we augment the MetRxn database with a new dataset of elementally balanced reaction operators using the automated atom mapping based reaction rule extraction procedure termed rePrime. For each reaction, the rePrime procedure identifies and captures as reaction rules the molecular graph topological changes underpinning the substrate to product graph conversion. A reaction rule captures the location of active reaction centers affected by the conversion of substrates to products. The reaction rules and known reactions are then operated upon a mixed integer linear programming algorithm (novoStoic) to identify a mass-balanced biochemical network that converts a source to a target metabolite while minimizing the number of*de novo* steps. We demonstrate the application of novoStoic in a few case studies such as designing shorter pathways from ferulate to protocatechuate and exploring Cð□›¼-Cð□›½ cleavage pathway of guaiacylglycerol-ð□›½-guaiacyl ether. By exploring the uncharted chemical space enabled by enzyme promiscuity, novoStoic paves the way to identify more carbon/energy efficient lignin funneling pathways with minimal heterologous enzymes to engineer, while exploring the organismâ€™s potential underground metabolism.

References

[1] Kumar, A., et al. Pathway design using de novo steps through uncharted biochemical spaces. *Nature communications* 9.1 (2018): 184.

Experimental Design for Parameter Estimation in Kinetic Models of Metabolism

Authors:

Mahadevan, R., University of Toronto
Euler, C., University of Toronto
Srinivasan, S., University of Toronto
Cluett, W., University of Toronto

In kinetic models of metabolism, the parameter values determine the dynamic behaviour predicted by these models. Estimating parameters from in vivo experimental data requires the parameters to be structurally identifiable, and the data to be informative enough to estimate these parameters. Existing methods to determine the structural identifiability of parameters in kinetic models of metabolism can only be applied to models of small metabolic networks due to their computational complexity. Additionally, a priori experimental design, a necessity to obtain informative data for parameter estimation, also does not account for using steady state data to estimate parameters in kinetic models. We present a scalable methodology to structurally identify parameters for each flux in a kinetic model of metabolism based on the availability of steady state data. We determine the number and nature of experiments for generating steady state data to estimate the enzyme kinetic parameters in a kinetic model of a small metabolic network. We show that most parameters in fluxes expressed by mechanistic enzyme kinetic rate laws (Michaelis-Menten, Monod-Wyman-Changeux and Hill kinetics) can be identified using steady state data. We also show that the requisite steady state data can be obtained from experiments involving both substrate and enzyme level perturbations. While substrate perturbation experiments are necessary for estimating parameters of uptake fluxes, enzyme level perturbation experiments may be necessary to estimate parameters of intracellular fluxes. The proposed methodology can be used in combination with other identifiability and experimental design algorithms that use dynamic data in order to determine the most informative experiments requiring the least resources to perform.

Systems Biology of Yeast Metabolism

Authors:

Nielsen, J., Chalmers University of Technology

Jens Nielsen[1,2]

[1] *Department of Biology and Biological Engineering, Chalmers University of Technology, Gothenburg, Sweden*

[2] *Novo Nordisk Foundation Center for Biosustainability, Technical University of Denmark, DK-2800 Lyngby, Denmark*

Metabolic Engineering relies on the Design-Build-Test cycle. This cycle includes technologies like mathematical modeling of metabolism, genome editing and advanced tools for phenotypic characterization. In recent years there have been advances in several of these technologies, which has enabled faster development of metabolically engineered strains that can be used for production of fuels and chemicals, but it is still challenging to perform efficient design. There is therefore in particular a need for advancing our ability to model metabolism, and this can be achieved through integration of systems biology tools. The yeast*Saccharomyces cerevisiae* is widely used for production of fuels, chemicals, pharmaceuticals and materials. Through metabolic engineering of this yeast a number of novel industrial processes have been developed over the last 10 years. Besides its wide industrial use, *S. cerevisiae* also serves as an eukaryal model organism, and many systems biology tools have therefore been developed for this organism. These tools can be used for detailed phenotypic characterization as well as for metabolic design. In this lecture, advances in metabolic modeling of yeast will be presented. It will be shown how integration of kinetic information into genome-scale metabolic models can significantly improve their predictive strength. Furthermore, through expanding these models to describe protein synthesis and translocation, it is possible to get new insight into what is constraining flux through different metabolic pathways. Finally, it will be shown how advances in quantitative omics allows us to provide a global map of flux control in the metabolic network.

Essential Metabolism for a Minimal Cell

Authors:

Breuer, M., University of Illinois at Urbana-Champaign

The question of the core requirements of cellular life led to the construction of the "minimal cell" JCVI-syn3A: A cell where practically all genes were removed that were not essential for robust growth in a stress-free laboratory environment. With only 493 genes in a 543 kbp genome, JCVI-syn3A has a genome smaller than that of any independently-replicating cell found in nature. It provides a versatile platform to study the basics of cellular life and is small enough that a complete description of all cellular functions can be pursued.
Here, we present an extensively curated metabolic reconstruction and flux balance analysis (FBA) model of this minimal cell, using the vast amount of experimental information available on its natural precursor,*Mycoplasma mycoides capri*. The model, featuring 337 reactions involving 304 metabolites, is near-complete with 98 % of enzymatic reactions justified through gene assignments and/or experimental evidence, and agrees well with gene essentiality data from transposon mutagenesis experiments. The 155 genes included in the reconstruction have a high *in vivo* essentiality or quasi-essentiality of 92 %, compared to 79 % *in silico* essentialityâ€“underscoring the minimality of the network. The reconstruction itself and the comparison of *in vivo* and *in silico*essentialities lead to new hypotheses on particular metabolic functions, suggesting specific experiments. Thus, the model provides a solid foundation for further experimental and computational studies on the minimal cell.

Characterization and Design of Phase Spaces and Yield Spaces in Genome-Scale Metabolic Models

Authors:

Zanghellini, J., University of Natural Resources and Life Sciences
Klamt, S., Max Planck Institute for Dynamics of Complex Technical Systems
Müller, S., University of Vienna
Regensburger, G., Johannes Kepler University Linz

Production rates and yields are key parameters of biochemical transformation processes. While optimal rates and phase spaces are readily studied with flux-balance analysis (FBA) approaches, optimal yields and yield spaces are rarely systematically analyzed. Often elementary flux modes (EFMs) are used to characterize yields and yield-optimal pathways. However, EFMs characterize the unbounded flux cone and are incompatible with non-zero flux bounds and allocation constraints often used in FBA.

By resorting to the concept of elementary flux vectors (EFVs), it is possible to generalize the idea of unique metabolic pathways to also account for inhomogeneous linear constraints. We show that any rate-optimal FBA solution sits in an optimal polyhedron spanned by (certain) EFVs. This holds true not only for rate-optimal but for yield-optimal solutions too, which cannot be found by standard FBA approaches.

Next, we demonstrate that (optimal) yield spaces can be readily calculated even in genome-scale metabolic models by linear-fractional programing without explicitly enumerating EFVs. Although phase spaces and yield spaces often are of similar shapes (and therefore sometimes confused), they carry very different information. In a realistic analysis based on *E. coli*, we show how these complementary pieces of information can be used to understand and optimally shape the metabolic capabilities of cell factories with any desired yield and/or rate requirements.

We conclude that EFVs provide an unifying framework for the theoretical description and analysis of any constraint-based model under arbitrary linear constraints. More specifically, EFVs close the gap between biased FBA approaches and unbiased EFM approaches and allow one to fully characterize and shape metabolic phase spaces and yield spaces. This reinforces the fundamental importance of EFVs (or EFMs) as the ``coordinates of metabolism´´. However, an explicit enumeration of EFVs is not required as phase spaces and yield spaces can be efficiently computed even in genome-scale metabolic networks.

DD-Decaf: Data-Driven Design of Cell Factories and Communities

Authors:

Sonnenschein, N., Technical University of Denmark
Kutuzova, S., Technical University of Denmark
Beber, M. E., Technical University of Denmark
Regestig, H., Technical University of Denmark
Herrgård, M. J., Technical University of Denmark
Kaafarani, A., Technical University of Denmark
Lopez, A., Technical University of Denmark
Dannaher, D., The Novo Nordisk Foundation Center for Biosustainability
Lieven, C., Technical University of Denmark
Fodor, M., Genialis, Inc

With ultra-precise genome editing tools at our disposal, we will see a transformation of the life sciences. We will shift away from experiments manipulating one factor at a time towards simultaneous changes at multiple loci. The success of those increasingly complex experiments will crucially depend on our ability to accurately predict systemâ€™s level behavior and will often require non-intuitive designs.

In principle, integrating omics data with systems biology models provides the means for optimizing knowledge gain through rational target selection and experimental design. This combination of methods is not leveraged effectively, however, due to a lack of readily available tools that allow us to rapidly access and analyze public and private data.

With this project, we aim to make a broad spectrum of omics data useful for biotechnology and life science research by integrating systems biology with design in a one-stop source.

A group of five renowned academic partners (DTU, Chalmers, EMBL, EPFL and UMinho) is driving research on integrative model-based omics data analysis to enable:

1. Metagenomics-enabled design of novel enzymes and biochemical pathways.
2. Omics data-driven design of cell factories for the production of chemicals and proteins.
3. Analysis and design of microbial communities relevant to human health, industrial biotechnology and agriculture.

All research efforts are integrated in an interactive, web-based platform available at http://dd-decaf.eu.

Dynamic Enzyme-Cost Flux Balance Analysis (deFBA) Modelling for an Industrially Relevant Methanotroph Methylomicrobium Buryatense

Authors:

De Becker, K., KU Leuven
Waldherr, S., KU Leuven

Constraint-based modelling has made significant progress in the last few years. Starting from metabolic network reconstructions a whole range of different modelling frameworks has been developed. The static flux balance analysis (FBA) method allows calculation of intracellular and exchange fluxes. Mahadevan et al. (2002) extended this approach to predict dynamic concentration profiles of extracellular metabolites, the so-called dynamic flux balance analysis method (DFBA). Recently, Waldherr et al. (2015) developed the dynamic enzyme-cost flux balance analysis (deFBA) framework. This approach includes enzyme production and takes into account that reaction fluxes are limited by the amount of catalyzing enzyme present inside the cells.

In this contribution, the construction of a deFBA model for the methanotroph *Methylomicrobium buryatense* is discussed. *M. buryatense* consumes methane and produces small organic acids. The deFBA model is built starting from a metabolic network reconstruction from de la Torre et al. (2015). Enzyme composition and catalytic information are obtained from online databases such as BRENDA, Kegg,

GenBank and Uniprot. A deFBA model is obtained containing 546 reactions from which 149 are enzyme production reactions. The model simulations predict specific substrate uptake and growth rates comparable to experimentally obtained values by Gilman et al. (2015). An exponential growth profile for biomass and its components is predicted. Production fluxes for acetic and formic acid are non-zero. However, lactic acid is not produced which contradicts experimental measurements.

Mahadevan et al., Dynamic flux balance analysis of diauxic growth in*Escherichia coli*, Biophys J., 2002;83(3):1331-1340.
Waldherr et al., Dynamic optimization of metabolic networks coupled with gene expression, Journal of Theoretical Biology, 2015;365:469-485

de la Torre et al., Genome-scale metabolic reconstructions and theoretical investigation of methane conversion in *Methylomicrobium buryatense* strain 5G(B1), Microbial Cell Fact., 2015;14:188
Gilman et al., Bioreactor performance parameters for an industrially-promising methanotroph*Methylomicrobium buryatense* 5GB1, Microbial Cell Fact., 2015;14:182

Quantifying Flux Uncertainty at the Genome Scale

Authors:

Backman, T., Joint BioEnergy Institute

Constraint-based metabolic flux analysis of knockout strategies is an efficient method to simulate the production of useful metabolites in microbes. Owing to the recent development of technologies for artificial DNA synthesis, it may become important in the near future to mathematically design minimum metabolic networks to simulate metabolite production. Accordingly, we have developed a computational method where parsimonious metabolic flux distribution is computed for designated constraints on growth and production rates which are represented by grids. When the growth rate of this obtained parsimonious metabolic network is maximized, higher production rates compared to those noted using existing methods are observed for many target metabolites. The set of reactions used in this parsimonious flux distribution consists of reactions included in the original genome scale model iAF1260. The computational experiments show that the grid size affects the obtained production rates. Under the conditions that the growth rate is maximized and the minimum cases of flux variability analysis are considered, the developed method produced more than 90% of metabolites, while the existing methods produced less than 50%. Mathematical explanations using examples are provided to demonstrate potential reasons for the ability of the proposed algorithm to identify design strategies that the existing methods could not identify. The source code is freely available, and is implemented in MATLAB and COBRA toolbox. This study has been accepted by BMC Bioinformatics.

Data Fusion Based on Coupled Matrix and Tensor Factorizations

Authors:

Acar, E., Simula Metropolitan Center for Digital Engineering

When the goal is to discover the underlying patterns in a complex system such as the human metabolism or brain, the complexity of the problem requires the collection and analysis of data from multiple sources. Therefore, data fusion, i.e., knowledge extraction by jointly analyzing complementary data sets, is a topic of interest in many fields. For instance, in metabolomics, analytical platforms such as Liquid Chromatography - Mass Spectrometry and Nuclear Magnetic Resonance Spectroscopy are used for chemical profiling of biological samples. Measurements from different platforms are capable of detecting different chemical compounds with different levels of sensitivity, and their fusion has the potential to provide a more complete picture of the metabolome related to a specific condition. Similarly, neuroimaging modalities such as functional Magnetic Resonance Imaging (fMRI) and electroencephalography (EEG) provide information about the brain function in complementary spatio-temporal resolutions, and their joint analysis is expected to provide better understanding of brain activities. However, data fusion remains a challenging task since there is a lack of data mining tools that can jointly analyze incomplete (i.e., with missing entries) heterogeneous (i.e., in the form of higher-order tensors and matrices) data sets, and capture the underlying shared/unshared patterns. We formulate data fusion as a coupled matrix and tensor factorization (CMTF) problem and discuss its extension to structure-revealing data fusion, i.e., fusion models that can identify shared and unshared factors in coupled data sets. Numerical experiments on real coupled data sets demonstrate that while traditional methods based on matrix factorizations have limitations in terms of jointly analyzing heterogeneous data sets, the structure-revealing CMTF model can successfully capture the underlying patterns by exploiting the low-rank structure of higher-order tensors. We will show the broad impact of CMTF-based fusion models with applications from metabolomics, neuroscience and recommender systems.

COBRA and Machine Learning Approaches for Engineering Microbial Biocatalysts

Authors:

Reed, J., University of Wisconsin Madison

Microbes have been engineered to produce a variety of chemicals, including biofuels, commodity chemicals, specialty chemicals, and therapeutics. Chemical production can be enhanced by connecting synthesis pathways to host metabolism, re-wiring regulatory networks, improving precursor production, and optimizing gene expression. A number of computational systems biology approaches have been developed to facilitate metabolic engineering efforts by suggesting which combination of genetic changes would improve chemical production. Network analysis methods can be used to identify paths from renewable substrates to high-value chemical products and central metabolic precursors common to a variety of chemical products. Genome-scale metabolic models can be used to predict how gene deletions, gene additions, and flux changes would impact chemical product yields, growth rates, and/or productivities. Additionally, machine learning and active learning algorithms can be used to optimize gene expression constructs to efficiently convert metabolic precursors into desired products. Case studies will be presented that show how a variety of computational tools can guide development of strains with enhanced chemical production. This work will illustrate how integrating computational and experimental efforts can lead to the rapid development of microbial biocatalysts for renewable chemical production.

How Accurate Is Automated Gap Filling of Metabolic Models?

Authors:

Karp, P., SRI International
Latendresse, M., SRI International

Reaction gap filling is a computational technique for proposing the addition of reactions to genome-scale metabolic models to permit those models to run correctly. Gap filling completes a reaction network by adding reactions that enable biosynthesis of all required metabolic products from available nutrients. The models are incomplete because they are derived from annotated genomes in which not all enzymes have been identified.

We present two studies of gap-filling accuracy. In the first study [1] we compared the results of applying an automated likelihood-based gap filler (MetaFlux) within the Pathway Tools software with the results of manually gap filling the same metabolic model. Both gap-filling exercises were applied to the same genome-derived metabolic reconstruction for *Bifidobacterium longum*. The MetaFlux gap filler attained recall of 61.5% and precision of 66.6%, taking the manual gap-filling result as the gold standard.

In the second study [2] we generated degraded versions of the EcoCyc-20.0-GEM model by randomly removing flux-carrying reactions from a growing model. We gap-filled the degraded models using 13 variations of MetaFlux (including the use of the SCIP and CPLEX Mixed Integer Linear Programming solvers and three different gap-filling algorithms) and compared the resulting gap-filled models with the original model. The best MetaFlux variation showed a best average precision of 87% and a best average recall of 61%. Although none of the 13 algorithm variations was best in all dimensions, we found one variation that was fast, accurate, and returned more information to the user. Some gap-filling variations were inaccurate, producing solutions that were non-minimum or invalid (did not enable model growth).

[1] "How Accurate is Automated Gap Filling of Metabolic Models?" Karp PD., Latendresse M., Weaver DW., Submitted.

[2] "Evaluation of reaction gap-filling accuracy by randomization," Latendresse M., Karp PD., BMC Bioinformatics 2018 19(1):53.

Bridging the Gap between Structural Systems Biology and Constraint-Based Modeling: Tools for Defining the Functional Structural Proteome and Extracting Genome-Scale Structural Information for Metabolic and Macromolecular Expression Models

Authors:

Catoiu, E., University of California, San Diego
Mih, N., UCSD
Lachance, J. C., Université de Sherbrooke
Lloyd, C. J., University of California, San Diego
Kavvas, E., UCSD
Palsson, B. O., University of California, San Diego

Genome-scale models of metabolism and macromolecular expression (ME-models) have used structural information to identify properties of the structural proteome by mapping individual genes to a subunit within a 3D protein structure. While gene-mapping is a first step in structural systems biology, enzymes that catalyze biochemical reactions can be composed of multiple protein subunits from multiple genes. Thus, the proper gene-mapping of multi-subunit protein structures to form a â€œfunctionalâ€ structural proteome is not only crucial for the integration of structural data into genome-scale models, but also provides the basis for discovery and incorporation of novel structural properties for constraint development in ME-models. Here, we describe the enhancement of a structural systems biology software (ssbio) tools to define and to gather novel genome-scale properties of the functional structural proteome of *Escherichia coli.* Specifically, tools were developed to address the technical challenges of working with structural data, ensuring that structures incorporated into*ssbio* are in a format that reflects the multi-subunit architecture of the enzyme. Structure quality predictions were improved to identify and remove mis-folded protein regions found in homology models. An algorithm was developed to distinguish between multiple conformations of structures available in the Protein Databank (PDB), to recreate annotated gene-enzyme stoichiometry, to define the functional structural proteome of *E.coli,* and to predict incorrect enzyme annotations. Upon the identification of the functional proteome, novel angstrom-level structural properties including cross-sectional area of transmembrane enzymes and soluble volume of enzymes were calculated. Additionally, computational tools to map residue-specific information such as enzymatic domains, secondary structure and strain-specific sequence variation onto structural data were deployed to analyze in-house mutations of interest. Together, the development of the aforementioned structural systems biology tools allows for seamless integration of functionally relevant structural information into genome-scale models, potentially leading to a new set of constraints driving biological discovery.

Scalable Tools for Analyzing Steady-State Microbial Communities Using Standardized Genome-Scale Metabolic Models

Authors:

Chan, S. H. J., The Pennsylvania State University
Maranas, C., The Pennsylvania State University

Extending genome-scale metabolic modeling to microbial communities is a promising way to unravel community interactions. We present three tools that significantly improve the accuracy and feasibility of simulating community metabolism. First, little emphasis has been placed on the need to impose a time-averaged constant growth rate across all community members to ensure co-existence and stability. Without this constraint, the faster growing organism will ultimately displace other microbes. We introduced the SteadyCom framework for predicting community metabolism consistent with the steady-state requirement. SteadyCom is scalable to a large number of organisms and compatible with flux variability analysis (FVA). In contrast to flux balance analysis, SteadyCom can predict an abundance profile with good agreement to experimental gut microbiota.
Second, during simulating community metabolism, an unstandardized biomass reaction of any organisms that produces biomass with a molecular weight different from the standard 1 g/mmol introduces a systematic error. We developed a computational procedure for checking the biomass weight so as to eliminate the error. 42 out of 64 examined

models show >5% discrepancies in biomass weights. We demonstrated that biomass weight discrepancies could cause significant errors in the predicted community composition that are disproportionately larger than the biomass weight discrepancies. Microbes with underestimated biomass weights are overpredicted whereas microbes with overestimated biomass weights are underpredicted.

Third, to effectively perform FVA in the absence of thermodynamically infeasible cycles in community models, we devised a method termed localized loop-less FVA (lll-FVA) with superior computational performance. We identified the fewest constraints required under the loop-less requirement and put forth the concept of localized loop-less constraints. The computational time is reduced by 10-150 times compared to the original loop-less constraints and by 4-20 times compared to the currently fastest method Fast-SNP. lll-FVA offers a scalable strategy for loopless flux calculations for large multi-compartment/multi-organism models (e.g., $>10^4$ reactions).

Genome-Scale Metabolic Model Embedding for Fed-Batch Optimal Feed Policy Determination

Authors:

Conejeros, R., Pontificia Universidad Catolica de Valparaiso
Scott, F., Universidad de los Andes
Vassiliadis, V., University of Cambridge

A new method for genome-scale metabolic model into culture kinetics, resulting in a dynamic flux balance analysis model is described and used to calculate optimal feed policies for fed-batch cultures where the dynamics of the fermentation process is captured by a set of differential equations and the metabolism is described by a genome scale flux balance analysis.

The methodology transforms the bounds of the embedded linear programming problems of flux balance analysis via a logarithmic barrier (interior point) approach. Further algebraic manipulations produce an implicit ordinary differential equations system.

The system of differential equations obtained, is embedded in an optimal control solver (using gPROMS), to optimize the feed flow rate profiles, substrate concentration in the feed and dissolved oxygen profiles in the culture. For genome-scale models, the optimal control problem has thousands of ordinary differential equations. Despite its size, a simulation can be performed in less than one second, and the full optimal control task can be completed in a fraction of a minute. This represents a significant improvement from previous works presented in literature, where problem reformulation based in collocation results in a large non-linear programming problem, and where only a network describing the central metabolism was addressed using this methodology.

Case studies include genome-scale metabolic models for *Escherichia coli*where total biomass accumulation at the end of the fermentation was maximized and a case study for ethanol accumulation from xylose and glucose using the yeast*Scheffersomyces stipitis*.

An Ecology of Modeling Methods, Data, and Tools to Decipher Functional Delegation and Species Interactions within a Microbiome

Authors:

Henry, C. S., Argonne National Laboratory
Weisenhorn, P., Argonne National Laboratory
van der Lelie, D., Gusto Global
Jeffryes, J. G., Argonne National Laboratory
Chivian, D. C., Joint BioEnergy Institute
Faria, J. P., Argonne National Laboratory
Edirisinghe, J. N., Argonne National Laboratory
Liu, F., Argonne National Laboratory
Seaver, S. M. D., Argonne National Laboratory
Dudley, H., University of Colorado Boulder
Zhang, Q., Argonne National Laboratory
Sadkhin, B., Argonne National Laboratory
Gupta, N., Argonne National Laboratory
Gu, T., Argonne National Laboratory
Taylor, R. C., Pacific Northwest National Laboratory
Song, H. S., Pacific Northwest National Laboratory
Bernstein, H. C., Pacific Northwest National Laboratory
Zucker, J., Pacific Northwest National Laboratory
Lindemann, S. R., Pacific Northwest National Laboratory

Gilbert, J., Argonne National Laboratory
Cottingham, R., Oak Ridge National Laboratory
Arkin, A., University of California, Berkeley
The term *metagenome* was first used in the literature 20 years ago, and in that time, we have witnessed a revolution in our capacity to gather data, model, and understand microbiome systems in natural and engineered environments. Finally we have sufficient data available to support the validation and refinement of metabolic models for microbiome systems. Here, I describe an ecology of tools, pipelines, and data that we have developed to support the modeling and simulation of microbiome systems. One of the initial challenges in modeling a microbiome system is to obtain species genomes from metagenomic data. To aid with this challenge, the KBase team recently implemented a pipeline for metagenome assembly, binning, QC, and annotation comprised of available open-source tools (e.g. metaSPAdes, MaxBin2). Another challenge is to rapidly construct predictive genome-scale models from newly assembled and binned genomes. To address this challenge, we have developed publicly accessible tools to generate interoperable models for microbial, plant, and fungal genomes (with recent significant improvements to our microbial model reconstruction pipeline). Even with models, these community systems are typically under-determined, with species interactions difficult to identify from noise. For this, we developed tools to predict auxotrophy from genomic data, revealing exciting new insights into how microbiome systems evolve and build stable interconnections. We also developed tools and pipelines to support the mapping of models and microbiome systems in general to metabolomics data, applying network approaches to enable models to slice through noise. Finally, we recently developed a new approach that explores the balance of protein-resources that plays a significant role in the delegation of functions in microbiome systems. In my talk, I describe these tools and explore the results from their application to a variety of

microbiome systems, including a photo-autotrophic microbial mat, an electrosynthetic microbiome, and a microbiome on a built surface.

Thermodynamic Framework for Mutant Phenotype Prediction

Authors:

Oyetunde, T., Washington University in St. Louis
Tang, Y. J., Washington University

Accurate prediction of the metabolic flux redistribution within mutant strains is vital for directing rational strain design. The discrepancies between experimentally determined and computationally predicted flux distributions imply that the predictive techniques do not accurately capture the mechanisms of cellular regulation.

Previous mutant prediction algorithms have essentially two components: (1) a metric to characterize the cellâ€™s desired metabolic state (for example flux or gene expression profiles) and (2) a metric to describe the distance from the desired state (for example, Euclidean distance). The mutant flux profile is then computed as the closest possible to the wild type state subject to the constraints of genetic or environmental perturbations. Different algorithms have reported significant gains in accuracy by tweaking the definitions of the two metrics.

To further improve the fidelity of knockout predictions and subsequent computational strain design, we developed a thermodynamics-based method, RElative MEtabolite Patterns (REMEP). REMEP hypothesizes that the optimum metabolic state is reflected in the energetic requirements to sustain flux through each metabolite node, and thus cell fluxomes adapt to perturbations from a reference state by preserving relative pattern of metabolite energy flows. REMEP performs better than comparable algorithms across different experimental datasets for *E. coli* and*S.*

cerevisiae (in terms of lower root mean square errors and higher Pearsonâ€™s correlation coefficients). These improvements support the REMEP assumption that cellular mechanisms of response to genetic and environmental perturbations leaves signatures that can be inferred from thermodynamics-derived metabolite patterns. The findings provide a new paradigm for genotype to phenotype mapping and insights into microbial flux network plasticity.

Dynamic FBA with Global Constraints on Cellular Protein Fraction

Authors:

Almaas, E., NTNU - Norwegian University of Science and Technology
Karlsen, E., NTNU-Norwegian University of Science and Technology
Schulz, C., NTNU - Norwegian University of Science and Technology
Martyushenko, N., NTNU - Norwegian University of Science and Technology

In this presentation, I will first briefly discuss recently developed software in our group that aids in the process of model reconstruction and curation: (i) *AutoKEGGRec* [1], a COBRA toolbox pipline based on KEGG that creates first draft metabolic network reconstructions of single organisms, reconstructions based on a list of organisms, and finally a consolidated reconstruction for a list of organisms or strains. (ii)*ModelExplorer* [2], a visual software framework to assist the user in exploring and curating genome-scale metabolic models provided in the sbml format. ModelExplorer automatically visualizes a metabolic network as a graph with a distinct separation and delineation of cellular compartments. Additionally, ModelExplorer highlights reactions and species that are unable to carry flux or participate in flux generation (ie. blocked), with the use of several different consistency checking modes available. Using high-quality models resulting from curation efforts, it is possible to apply a plethora of constraint-based methods to analyze the models' phenotypes. I will discuss a recently developed method that combines dynamic Flux Balance Analysis (dFBA) with a global constraint on cellular protein fraction [3]. We apply our method to a recent *E. coli* genome-scale reconstruction, finding that our method outperforms existing dFBA formulation. The globally-constrained dFBA method makes it possible to investigate cellular proteome composition in response to changing nutrient conditions.

[1] E. Karlsen, C. Schulz and E. Almaas. Submitted.
[2] N. Martyushenko and E. Almaas. Submitted.
[3] E. Karlsen and E. Almaas. In preparation.

Incorporating Flux Sampling into a Minimal Assumption Dynamic Flux Balance Analysis Algorithm

Authors:

Wilken, S. E., UCSB
O'Malley, M., UCSB
Petzold, L., UCSB

Recent advances in sequencing technology have contributed to the rapid development of several highly predictive genome-scale models. Techniques like flux balance analysis (FBA) allow for relatively simple, mechanistically based, interrogation of these models. A drawback of FBA is the assumption of steady state. Dynamic flux balance analysis (DFBA) is an extension of this technique to non-steady state systems; a primary advantage of this approach is the low extra cost associated with developing dynamic models that are still mechanistically valid.

DFBA simulators typically solve successive FBA problems, and use the resultant fluxes to move the dynamic states of the system forward in time. A challenge associated with DFBA is the proper way of handling the degenerate solution space resulting from FBA. We propose a novel algorithm that takes advantage of the degenerate solutions of FBA to more closely match *in vivo*conditions. Indeed, it is well known that noisy gene expression results in variable enzyme concentrations between clonal cells; therefore, one expects the metabolic fluxes experienced by each cell to be different. Our algorithm models this behavior by sampling from the solution space, and uses these sampled fluxes to move the states of the system forward in aggregate. The resultant time varying flux distributions represent an extension of flux variability analysis to the dynamic setting. Our algorithm stands in contrast to the current state of the art simulators that utilize sequential linear optimization to find a unique flux for each state that is integrated. While this approach is mathematically robust, it requires further assumptions that might not be valid, or could be hard to justify. If the predicted transient flux distributions associated with our approach prove to accurately capture the *in vivo* behavior of an organism, it would allow experimentalists the ability to carefully interrogate proposed engineering strategies in the dynamic setting.

Creation and Analysis of Biochemical Constraint-Based Models: The Cobra Toolbox v3.0

Authors:

Fleming, R. M. T., Leiden University

COnstraint-Based Reconstruction and Analysis (COBRA) provides a molecular mechanistic framework for integrative analysis of experimental molecular systems biology data and quantitative prediction of physicochemically and biochemically feasible phenotypic states. The COBRA Toolbox is a comprehensive desktop software suite of interoperable COBRA methods. It has found widespread applications in biology, biomedicine, and biotechnology because its functions can be flexibly combined to implement tailored COBRA protocols for any biochemical network. This protocol is an update to the COBRA Toolbox 1.0 and 2.0. Version 3.0 includes new methods for quality controlled reconstruction, modelling, topological analysis, strain and experimental design, network visualisation as well as network integration of chemoinformatic, metabolomic, transcriptomic, proteomic, and thermochemical data. New multi-lingual code integration also enables an expansion in COBRA application scope via high-precision, high-performance, and nonlinear numerical optimisation solvers for multi-scale, multi-cellular and reaction kinetic modelling, respectively. This protocol overviews all of these new features and can be adapted to generate and analyse constraint-based models in a wide variety of scenarios. The COBRA Toolbox 3.0 provides an unparalleled depth of constraint-based reconstruction and analysis methods.

Memote: A Community Driven Effort Towards a Standardized Genome-Scale Metabolic Model Test Suite

Authors:

Lieven, C., Technical University of Denmark
Beber, M. E., Technical University of Denmark
Olivier, B. G., Systems Bioinformatics, Amsterdam Institute for Molecules, Medicines and Systems, Vrije Universiteit Amsterdam
Bergmann, F. T., Heidelberg University
Sonnenschein, N., Technical University of Denmark

Several studies have shown that neither the formal representation nor the functional requirements of genome-scale metabolic models (GEMs) are precisely defined. Without a consistent standard, comparability, reproducibility, and interoperability of models across groups and software tools cannot be guaranteed.

Here, we present memote (https://github.com/opencobra/memote) an open-source software containing a community-maintained, standardized set of metabolic model tests. The tests cover a range of aspects from annotations to conceptual integrity and can be extended to include experimental datasets for automatic model validation. In addition to testing a model once, memote can be configured to do so automatically, i.e., while building a GEM. A comprehensive report displays the modelâ€™s performance parameters, which supports informed model development and facilitates error detection.

Memote provides a measure for model quality that is consistent across reconstruction platforms and analysis software and simplifies collaboration within the community by establishing workflows for publicly hosted and version controlled models.

Pathogen Metabolism and Antibiotic Resistance

COBRA Distinguished Young Investigator Award Winner

Authors:

Chandrasekaran, S., University of Michigan - Ann Arbor

Microbes have been engineered to produce a variety of chemicals, including biofuels, commodity chemicals, specialty chemicals, and therapeutics. Chemical production can be enhanced by connecting synthesis pathways to host metabolism, re-wiring regulatory networks, improving precursor production, and optimizing gene expression. A number of computational systems biology approaches have been developed to facilitate metabolic engineering efforts by suggesting which combination of genetic changes would improve chemical production. Network analysis methods can be used to identify paths from renewable substrates to high-value chemical products and central metabolic precursors common to a variety of chemical products. Genome-scale metabolic models can be used to predict how gene deletions, gene additions, and flux changes would impact chemical product yields, growth rates, and/or productivities. Additionally, machine learning and active learning algorithms can be used to optimize gene expression constructs to efficiently convert metabolic precursors into desired products. Case studies will be presented that show how a variety of computational tools can guide development of strains with enhanced chemical production. This work will illustrate how integrating computational and experimental efforts can lead to the rapid development of microbial biocatalysts for renewable chemical production.

A Computational Knowledge-Base Elucidates the Response of Staphylococcus Aureus to Different Media Types

Authors:

Seif, Y., UCSD
Palsson, B. O., University of California, San Diego
Monk, J., UCSD
Mih, N., UCSD
Poudel, S., UCSD
Broddrick, J. T., University of California, San Diego
Zuniga P, C., University of California
Tsunemoto, H., University of California San Diego
Zengler, K., University of California
S. aureus is classified as a serious threat pathogen and is a priority that guides the discovery and development of new antibiotics. It has been observed to exhibit differential antibiotic resistance profiles when cultured in different media. Despite growing knowledge of *S. aureus* metabolic capabilities, our understanding of its systems-level responses to different media types remains incomplete. Here, we develop a manually reconstructed genome-scale model (GEM-PRO) of metabolism with 3D protein structures for *S. aureus* USA300 str. JE2 containing 854 genes, 1,440 reactions, 1,327 metabolites and 673 3-dimensional protein structures. Computations were in 85% agreement with gene essentiality data from random barcode transposon site

sequencing (RB-TnSeq) and 68% agreement with experimental physiological data. Comparison of computational predictions with experimental observations highlight: 1) cases of non-essential biomass precursors; 2) metal cofactor promiscuity; 3) genes subject to transcriptional regulation involved in Staphyloxanthin biosynthesis; and 4) gaps of knowledge in C2-carbon assimilation; 5) the essentiality of purine and amino acid biosynthesis in synthetic physiological medium; and 6) a switch to aerobic glycolysis upon exposure to extracellular glucose that was elucidated using a time-course of quantitative metabolomics data. An up-to-date GEM-PRO thus serves as a knowledge-based platform to elucidate *S. aureusâ€™* metabolic response to its environment.

A Deep Dive into False Positive Gene Essentiality Predictions with Modeling and Multi-Omics Analysis

Authors:

Feist, A. M., University of California, San Diego

Essentiality assays are commonly practiced as important tools for the discovery of gene functions. Growth/no growth screens of single gene knockout strain collections are often utilized to test the predictive power of COBRA models. False positive predictions occur when computational analysis predicts a gene to be non-essential, however experimental screens deem the gene to be essential. In this study, we explored the definition of conditional essentiality from a phenotypic and genomic perspective. Gene-deletion strains associated with false positive predictions of gene essentiality on defined minimal medium for *E. coli* were targeted for extended growth tests followed by population sequencing. Of the twenty false positive strains available and confirmed from the Keio knock-out collection, 11 strains were shown to grow with longer incubation periods making these actual true positives. These strains grew reproducibly with a diverse range of growth phenotypes. It was found that 9 out of 11 of the false positive strains that grew acquired mutations in at least one replicate experiment and the types of mutations ranged from SNPs and small indels associated with regulatory or metabolic elements to large regions of genome duplication. Comparison of the detected adaptive mutations and modeling predictions of alternate pathways and isozymes suggested agreement for the observed growth phenotype for 6 out of the 9 cases where mutations were observed, which was further validated by analysis of the transcriptomes of selected strains. In conclusion, longer-term growth experiments followed by whole genome sequencing and transcriptome analysis can provide a better understanding of conditional gene essentiality and mechanisms of adaptation to such perturbations. Compensatory mutations are largely reproducible mechanisms and are in agreement with genome-scale modeling predictions to loss of function gene deletion events.

A Distance Measure for Heterogeneity Using Genome Scale Metabolic Networks

Authors:

Cabbia, A., Eindhoven University of Technology
Hilbers, P. A. J., Eindhoven University of Technology
van Riel, N. A. W., Eindhoven University of Technology

Physiological differences in the aging process are inherently present in a population, and increase with age, affecting the risk of developing disabilities and age-related diseases [1]. Patient-Derived Genome-Scale Metabolic Models (PD-GSMM) are built from human GSMM and experimental data, mostly transcriptomics and proteomics, belonging to single individuals. Personalized genome scale models have recently been used to plan individualized anti-cancer therapies [2], and to address the variability among cancer patients, identifying key genes involved in tumour growth [3]. Despite their success in cancer metabolism, is still not clear the extent to which PD-GSMs are representations of individual metabolic features in physiological conditions, and how successful such models are in capturing inter-individual heterogeneity when dealing with subtler phenotypes such as ageing. Starting from microarray datasets of younger and older adultsâ€™ skeletal muscle gene expression, we developed the first collection of patient-derived genome scale metabolic models of ageing individuals' myocytes, and used a data science approach to define a distance metric and assess the variability between metabolic models. This research is part of the PANINI project (**P**hysical **A**ctivity and **N**utrition**IN**fluences **i**n Aging), and has received funding from the

European Unionâ€™s Horizon2020 programme, under the Marie Sklodowska-Curie grant agreement 675003.
[1] Yang Y and Lee L.C.

Dynamics and Heterogeneity in the Process of Human Frailty and Aging: Evidence From the U.S. Older Adult Population

J Gerontol B Psychol Sci Soc Sci. 2010 Mar; doi: 10.1093/geronb/gbp102

[2] Agren R. et al.

Identification of anticancer drugs for hepatocellular carcinoma through personalized genome-scale metabolic modeling.

Mol Syst Biol. 2014 Mar; doi: 10.1002/msb.145122.

[3] Uhlen M. et al.

A pathology atlas of the human cancer transcriptome.

Science. 2017 Aug 18; doi: 10.1126/science.aan2507.

A Machine Learning-Empowered Kinetic Reconstruction of E. coli Metabolism

Authors:

Zielinski, D. C., University of California
Palsson, B. O., University of California - San Diego

Constraint-based models of metabolism have been powerful tools for characterizing and predicting organism function with minimal parameter requirements. However, several important applications of metabolic models have emerged that depend upon enzyme kinetic parameters, which are scarce in the literature. In this work, we present a genome-scale kinetic reconstruction of the *E. coli*metabolic network. We curated the literature and existing databases to extract available kinetic data, including enzyme kinetic parameters, kinetic mechanisms, and protein structural information. To address the data scarcity issue, we fill gaps in measured parameters with statistical models for kcat, Km, and Keq trained on literature data. We reconcile potential inconsistencies in data using a nonlinear regression approach that finds rate constants satisfying kinetic data for each enzyme to the extent possible. The resulting kinetic reconstruction should be a valuable resource to the community and empower parameter-dependent systems biology research.

A Metabolic Reconstruction of Lactobacillus Reuteri and Analysis of Its Potential As a Cell Factory

Authors:

Kristjansdottir, T., University of Iceland
Bosma, E. F., DTU Biosustain
Branco dos Santos, F., University of Amsterdam
Özdemir, E., Technical University of Denmark
Nielsen, A. T., Technical University of Denmark
Gudmundsson, S., University Of Iceland

Lactic acid bacteria (LAB) have long been used by humans in the food industry and is*Lactobacillus reuteri* one of them. Additionally, *L. reuteri* is also known to produce several different compounds, with potential biotechnological interest, such as; the antimicrobial compound reuterin, vitamin B12 and 1,3-propanediol, which can be used as a starting material for plastic production. In an attempt to further explore*L. reuteri* as a cell factory for production of valuable compounds, a genome-scale metabolic model was reconstructed and thoroughly curated and validated using data from the literature and several experimental data sets, including knockout strains. The biomass reaction was built from species specific data when available and alternatively from *L. plantarum* and *L. lactis*.

L. reuteri metabolizes glucose via the phosphoketolase pathway. Genome data indicates the absence of a phosphofruktokinase (PFK), which results in an inactive EMP pathway. Model simulations show that PFK does not provide a significant advantage in terms of growth.

A pathway prediction algorithm was used to identify several heterologous pathways for production of 1-propanol. Theoretical yields of 1-propanol were evaluated for multiple carbon sources. The most promising pathways were further optimized *in silico* by identifying targets for overexpression and knockouts for increased production of 1-propanol.

A Model of Cardiomyocyte Metabolism

When the goal is to discover the underlying patterns in a complex system such as the human metabolism or brain, the complexity of the problem requires the collection and analysis of data from multiple sources. Therefore, data fusion, i.e., knowledge extraction by jointly analyzing complementary data sets, is a topic of interest in many fields. For instance, in metabolomics, analytical platforms such as Liquid Chromatography - Mass Spectrometry and Nuclear Magnetic Resonance Spectroscopy are used for chemical profiling of biological samples. Measurements from different platforms are capable of detecting different chemical compounds with different levels of sensitivity, and their fusion has the potential to provide a more complete picture of the metabolome related to a specific condition. Similarly, neuroimaging modalities such as functional Magnetic Resonance Imaging (fMRI) and electroencephalography (EEG) provide information about the brain function in complementary spatio-temporal resolutions, and their joint analysis is expected to provide better understanding of brain activities. However, data fusion remains a challenging task since there is a lack of data mining tools that can jointly analyze incomplete (i.e., with missing entries) heterogeneous (i.e., in the form of higher-order tensors and matrices) data sets, and capture the underlying shared/unshared patterns. We formulate data fusion as a coupled matrix and tensor factorization (CMTF) problem and discuss its extension to structure-revealing data fusion, i.e., fusion models that can identify shared and unshared factors in coupled data sets. Numerical experiments on real coupled data sets demonstrate that while traditional methods based on matrix factorizations have limitations in terms of jointly analyzing heterogeneous data sets, the structure-revealing CMTF model can successfully capture the underlying patterns by exploiting the low-rank structure of higher-order tensors. We will show the broad impact of CMTF-based fusion models with applications from metabolomics, neuroscience and recommender systems.

A Multi-Omics Investigation Unveiling the Tiered Regulation of Breast Cancer Cell Metabolism

Authors:

Katzir, R., University of Maryland

Altered metabolism is a hallmark of cancer, but little is still known about its regulation. In this study, we measure transcriptomic, proteomic, phospho-proteomic and fluxomics data in a breast cancer cell-line (MCF7) across three different growth conditions. Integrating these multiomics data within a genome scale human metabolic model in combination with machine learning we systematically chart the different layers of metabolic regulation in breast cancer cells, predicting which enzymes and pathways are regulated at which level. We distinguish between two types of reactions, directly or indirectly regulated. Directly-regulated reactions include those whose flux is regulated by transcriptomic alterations (~890) or via proteomic or phospho-proteomics alterations (~140) in the enzymes catalyzing them. Potentially indirectly regulated reactions are those that currently lack evidence for direct regulation (~930). Many metabolic pathways are predicted to be regulated at different levels, and those may change at different media conditions. Remarkably, we find that the flux of predicted indirectly regulated reactions is strongly coupled to the flux of the predicted directly regulated ones, uncovering a hierarchical organization of breast cancer cell metabolism. Furthermore, the predicted indirectly regulated reactions are predominantly reversible. Taken together, this architecture may facilitate the rapid and efficient metabolic reprogramming in response to the varying environmental conditions incurred by the tumor cells. The approach presented lays a conceptual and computational basis for a more complete mapping of metabolic regulation in different cancers with incoming additional data.

A Multiobjective Strain Design Platform for Modular Cell Engineering

Authors:

Garcia, S., University of Tennessee
Trinh, C. T., University of Tennessee

Metabolic engineering has enabled the use of microbial cell factories for industrial production of biochemicals. However, developing an optimal strain for synthesis of one product with the conventional strategy is laborious and expensive. To accelerate and reduce the cost of strain engineering, we formulate the modular cell (MODCELL) design principles by exploiting the modular organization of metabolic networks and combinatorial possibilities of metabolic modules that enable the synthesis of a large space of biochemicals. Using multiobjective optimization methods, we develop novel algorithms to implement the MODCELL design for genome-scale metabolic networks and the associated software ModCell2.0. We demonstrate ModCell2.0 for design, construction, and validation of an *E. coli*modular cell for combinatorial synthesis of biochemicals, e.g., alcohols and bioesters from fermentable sugars and organic wastes. We envision MODCELL will provide a useful platform for modular cell engineering.

A Systematic Assessment of Current Genome-Scale Metabolic Reconstruction Tools

Authors:

Mendoza, S. N., Systems Bioinformatics, Amsterdam Institute for Molecules, Medicines and Systems, VU Amsterdam, The Netherlands
Olivier, B. G., Systems Bioinformatics, Amsterdam Institute for Molecules, Medicines and Systems, Vrije Universiteit Amsterdam
Teusink, B., Systems Bioinformatics, Amsterdam Institute for Molecules, Medicines and Systems, VU Amsterdam, The Netherlands
Molenaar, D., Systems Bioinformatics, Amsterdam Institute for Molecules, Medicines and Systems, Vrije Universiteit Amsterdam
Several genome-scale metabolic reconstruction software platforms have been developed and continuously updated during the last fifteen years. These tools have been widely applied to reconstruct metabolic models for hundreds of microorganisms ranging from important human pathogens to species of industrial relevance. So far, these platforms have not been systematically evaluated with respect to software quality, best potential uses and intrinsic capacity to generate high-quality genome-scale metabolic models. In consequence, this lack of assessment keeps users away from selecting the tool that best fits the purpose of their research. In this work, we reviewed the current genome-scale reconstruction software platforms and we performed a systematic assessment of the most promising tools. We defined a list of features for assessing software quality related to genome-scale reconstruction which we hope it would be useful for potential users. In addition, we compared the output networks

generated using each of these tools with the high-quality manually curated models of*Lactobacillus plantarum* and *Bortedella pertussis*, representatives of gram-positive and gram-negative bacteria, respectively. We showed that none of the tools outperform the others in all the studied features and that users should carefully choose one or another, or even combining them, depending on the particularities of their projects.

An Enhanced Genome-Scale Metabolic Reconstruction of Streptomyces Clavuligerus Identifies Novel Strain Improvement Strategies

Authors:

Rios-Estepa, R., University of Antioquia

Toro, L., Universidad de Antioquia

Pinilla-Mendoza, L., Universidad de Antioquia

Avignone-Rossa, C., University of Surrey

Extending genome-scale metabolic modeling to microbial communities is a promising way to unravel community interactions. We present three tools that significantly improve the accuracy and feasibility of simulating community metabolism. First, little emphasis has been placed on the need to impose a time-averaged constant growth rate across all community members to ensure co-existence and stability. Without this constraint, the faster growing organism will ultimately displace other microbes. We introduced the SteadyCom framework for predicting community metabolism consistent with the steady-state requirement. SteadyCom is scalable to a large number of organisms and compatible with flux variability analysis (FVA). In contrast to flux balance analysis, SteadyCom can predict an abundance profile with good agreement to experimental gut microbiota.

Second, during simulating community metabolism, an unstandardized biomass reaction of any organisms that produces biomass with a molecular weight different from the standard 1 g/mmol introduces a systematic error. We developed a computational procedure for checking the biomass weight so as to eliminate the error. 42 out of 64 examined models show >5% discrepancies in biomass weights. We demonstrated that biomass weight discrepancies could cause significant errors in the predicted community composition that are disproportionately larger than the biomass weight discrepancies. Microbes with underestimated biomass weights are overpredicted whereas microbes with overestimated biomass weights are underpredicted.

Third, to effectively perform FVA in the absence of thermodynamically infeasible cycles in community models, we devised a method termed localized loop-less FVA (lll-FVA) with superior computational performance. We identified the fewest constraints required under the loop-less requirement and put forth the concept of localized loop-less constraints. The computational time is reduced by 10-150 times compared to the original loop-less constraints and by 4-20 times compared to the currently fastest method Fast-SNP. lll-FVA offers a scalable strategy for loopless flux calculations for large multi-compartment/multi-organism models (e.g., >10^4 reactions).

An Integrated COBRA-PBPK Model to Study Interactions between the Gut Microbiome and the Brain in Autism

Authors:

Palukuri, M., University of Texas at Austin
Shivakumar, S., Georgia Institute of Technology
Sahoo, S., Indian Institute of Technology Madras
Rengaswamy, R., Indian Institute of Technology Madras

Autism, also ASD (Autism Spectrum Disorder) is a complex neurological condition characterized by social, behavioural and developmental issues such as impaired communication and repetitive actions. Gastro-intestinal disturbances in autistic patients have motivated experiments that identified abnormal concentrations of certain bacteria in the gut. To understand the role played by the most reported gut bacteria associated with autism, namely *Bacteroides vulgatus, Clostridium perfringens, Desulfovibrio desulfuricans, Bifidobacterium longum* and *Lactobacillus acidophilus*, a hybrid modeling approach was adopted. Genome-scale metabolic models of these bacteria were integrated with the human small intestinal enterocyte model. Simultaneously, a permeability-limited, two sub-compartment, whole body PBPK transport model was incorporated to determine the distribution of gut-bacteria derived toxins throughout organs. A novel framework was formulated for the development and analysis of an integrated gut-PBPK-brain model and was applied to the study of autism for the first time. The integrated model predicted that intake of probiotics (*L. acidophilus* and *B. longum*) leads to reduced oxidative stress in the brain, resulting in relief from autistic symptoms.

Further, investigating the effect of various dietary schemes indicated that a typical western diet worsens symptoms in autistic individuals. Importantly, potential biomarkers for autism were predicted in terms of bacterial secretion products and affected pathways in the genome-scale metabolic model of the brain. Thus, the integrated model breaks new ground in developing a quantitative model for studying gut-brain interactions, with the potential to understand and remedy several neurological problems with links to the gut, such as autism, depression and anxiety. More generally, the framework for integration and analysis of constraint-based and PBPK models can be applied to study dynamically, the interactions between systems in various fields. Its applications are demonstrated by providing a mechanistic basis for autism pathogenesis and by proposing dietary changes and intake of probiotics for the alleviation of autistic symptoms.

Automated Reconstruction, Evaluation and Comparison of Diverse Genome Scale Fungal Models

Authors:

Edirisinghe, J. N., Argonne National Laboratory
Faria, J. P., Argonne National Laboratory
Liu, F., Argonne National Laboratory
Calhoun, S., Lawrence Berkeley National Laboratory
Grigoriev, I., Lawrence Berkeley National Laboratory
Henry, C. S., Argonne National Laboratory

Genome-scale metabolic models offer an efficient mechanism for predicting phenotypes across many environmental conditions. These models are also key tools in understanding interactions between species in the environment, including microbial, plant, and fungal species. While methods have emerged for the automated reconstruction of microbial and plant models from genomic data, the construction of high-quality fungal models remains a challenge. Here we introduce a new method for constructing genome-scale fungal models in an automated fashion based on a curated set of reactions and gene associations that are derived from 14 published fungal metabolic models. As the basis for our method, we produced a fungal model template that encompasses the biochemistry data from the published fungal models and the structural annotations from the associated fungal genomes. Using structural annotations of a fungal genome we compute a set of orthologous proteins against the curated fungal template in order to assert the presence or absence of specific biochemical reactions and pathways. Once the orthologous protein families are determined, the related biochemistry data is propagated to construct a new draft metabolic model. This method is deployed in the KBase (https://narrative.kbase.us/) as an app called "Build Fungal Model". We have built tools in KBase to produce high-quality metabolic models on prokaryotes and plants and new approaches to identify trophic interactions between species. Using these methods, we are able to study how fungal-bacterial, plant-fungal and microbiomes evolve as a sustainable community.

We applied â€œBuild Fungal Modelâ€ app to build draft models for more than 100 fungal genomes imported from the Joint Genome Institute (JGI) MycoCosm resource (https://genome.jgi.doe.gov/programs/fungi/index.jsf).

We compared these models side-by-side, exploring how each genome overlaps with our curated model template and plotting model variance along the phylogenetic tree of fungal genomes. All draft fungal models are available for download via a KBase Narrative.

Bayesian Inference of Metabolic Kinetics from Genome-Scale Multiomics Data

Authors:

St. John, P. C., National Renewable Energy Laboratory
Strutz, J., Northwestern University
Bomble, Y. J., National Renewable Energy Laboratory

Optimizing the metabolism of microorganisms for maximum yields and titers is a critical step in improving bioprocess economics. With the growing availability of transcriptomic, proteomic, metabolomic, and fluxomic analysis techniques, characterization of engineered strains has become increasingly detailed. However, utilizing multiomics data to make informed decisions about future strain improvements remains a major challenge in modern bioengineering. Parameterizing kinetic models from indirect, *in vivo* data is typically infeasible at the genome-scale. There is therefore a need for mechanistic modeling frameworks that can consume the large amount of data generated through multiomics experiments to yield actionable insight for strain engineers.

Metabolic ensemble modeling has emerged in recent years as an effective tool for estimating confidence intervals in kinetic metabolic models from observable steady-state flux and concentration measurements. The method operates by sampling feasible kinetic parameters and filtering to sets of parameters that match experimental observations. However, these methods rely on an explicit numerical integration of large kinetic models, and therefore scale poorly both in the size of the network that can be modeled, as well as the amount of data that can be used to constrain the parameters.

We demonstrate the coupling of linear-logarithmic kinetics to a Bayesian inference framework. The resulting method therefore represents a scalable, flexible method for modeling omics data. Specifically, linear-logarithmic kinetics enable steady-state fluxes to be predicted linearly from kinetic parameters, removing the computational burden associated with solving for steady-state flux. Additionally, since linear solutions permit the easy determination of likelihood gradients, advanced Bayesian inference techniques can be applied to reliably estimate posterior parameter distributions even for high-dimensional models. We demonstrate the method through a number of case studies, from simple *in vitro* reaction systems to complex, medium-scale metabolic models with hundreds of metabolites and reactions parameterized with modern multi-omic data.

Bofdat: Generating Biomass Objective Function for Constraint-Based Metabolic Models from Experimental Data

Authors:

Lachance, J. C., Université de Sherbrooke

The computation of growth phenotypes by genome-scale metabolic models (GEMs) requires the definition of an objective function. To simulate growth related phenotypes a biomass objective function (BOF) that includes key biomass components of the cell such as the major macromolecules (DNA, RNA, proteins), lipid composition, coenzymes, inorganic ions and specie-specific components. While the accurate definition of the BOF was shown to provide precise phenotypic predictions, no standardized computational platform is available to generate species-specific data-driven BOFs. To fulfill this gap in the software ecosystem, we implemented BOFdat: a Python package for the definition of **B**iomass **O**bjective**F**unction from experimental **dat**a. BOFdat is a 3-step process. At each step, different macromolecular categories are added along with their macromolecular weight fractions, facilitating the mass balancing of the objective function. The first step adds the main macromolecules and lipids using both omics data and macromolecular weight fraction of each category to calculate the stoichiometric coefficients of each metabolite. The second step determines coenzymes by performing a connectivity analysis of the entire metabolic network, and inorganic ions by comparing against a list of ions found in all previously generated metabolic models in the BiGG Models database. The third step is a novel, unbiased approach using a genetic algorithm to find the combination of metabolites that best match experimental gene essentiality data, and clusters the results into sets of metabolic end goals. We used BOFdat to reconstruct the BOF of the *Escherichia coli* model *i*ML1515 and found that the BOF generated by BOFdat had more similar biomass composition, growth rate, essentiality to the original predictions than other available methods. By providing an unbiased, data-driven approach to defining biomass objective functions, BOFdat has the potential to improve the quality of new genome-scale models and also greatly decrease the time required to generate a new reconstruction.

Characterizing and Ranking Computed Metabolic Engineering Strategies

Authors:

Schneider, P., Max Planck Institute for Dynamics of Complex Technical Systems
Klamt, S., Max Planck Institute for Dynamics of Complex Technical Systems

Computational strain design, i.e. the calculation of metabolic intervention strategies from a mathematical model, is a key component of an integrated metabolic engineering approach. A broad range of methods has been developed for this task, including bilevel optimization approaches (e.g.,OptKnock and RobustKnock) and the framework of Minimal Cut Sets (MCSs). Most of these algorithms enforce the coupling between growth and the synthesis of the desired product. Some of them, such as the MCSEnumerator, may return a large pool of thousands of possible intervention strategies from which the most suitable strategy can then be selected.

This work focuses on how to characterize and rank, in a meaningful way, a given pool of intervention strategies calculated for growth-coupled product synthesis. Some criteria are straightforward and can be assessed easily, for example, the number of necessary reaction or gene cuts, the maximal growth rate and the guaranteed (minimum) product yield. Additionally, we present more extensive criteria that are worth considering when picking the â€˜bestâ€™ intervention strategy. Among others we investigate the existence of metabolites that may disrupt the growth coupling when accumulated or secreted and check whether the interventions interrupt pathways at their origin or (less preferred) at later steps. We also assess the maximum thermodynamic driving force of the pathway(s) favored by the intervention strategy and take into account whether a specific intervention strategy relies on flux re-routing within the central metabolism or on other pathways with possibly minor throughputs. Furthermore, strategies that have significant overlap with alternative solutions are also ranked higher because they provide flexibility in implementation.

We applied the presented ranking approach to several sets of intervention strategies for growth-coupled synthesis of native and heterologous products computed from the genome-scale *E. coli* model iJO1366 using the MCSEnumerator algorithm.

COBRA.jl - Gearing up for the huge scale

Authors:

Heirendt, L., University of Luxembourg
Arreckx, S., University of Luxembourg
Satagopam, V. P., University of Luxembourg
Schneider, R., University of Luxembourg
Thiele, I., University of Luxembourg
Fleming, R. M. T., Leiden University

Ever larger biochemical networks are being developed by the COnstraint-Based Reconstruction and Analysis (COBRA) community. Reconstructions with millions of biochemical reactions are within reach, and standard implementations of COBRA software written in MATLAB, Python, or C, hit their limit. Existing implementations are not suitable for simultaneously analyzing thousands of these large networks and simulations are months-long.

Accelerating existing and new COBRA methods bears the challenges of short development times and speeding up analyses significantly, all while assuring scalability and parallelism across multiple computing nodes and without recurring to complex message-passing interfaces. The Julia language [1] satisfies this need. It is set to become the language of choice to accelerate analyses of large (< 500k biochemical reactions) and huge-scale biochemical networks (> 500k reactions).

A step towards gearing up for high-dimensional COBRA modeling is the release of the high-level, high-performance, and open-source COBRA.jl package (git.io/COBRA.jl). DistributedFBA.jl [2], part of COBRA.jl, allows performing a flux balance analysis and many related analysis types efficiently, especially on large and huge-scale models. Most COBRA analyses are based on the COBRA Toolbox [3], a comprehensive software suite of interoperable methods written in MATLAB, which has found widespread applications in biology, biomedicine, and biotechnology. PALM.jl allows launching analyses across several computing nodes simultaneously, and this for thousands of models at once.

Through the open-source high-performance COBRA.jl package, the analysis capabilities of the COBRA community are lifted to another level, as reconstruction and analysis of large and huge-scale models are made possible and accelerated.

[1] Bezanson et al., Julia: A Fresh Approach to Numerical Computing, SIAM Review, 59(1):65â€“98, 2017.
[2] Heirendt et al., DistributedFBA.jl: high-level, high-performance flux balance analysis in Julia, Bioinformatics, 33(9):1421â€“1423, 2017.
[3] Heirendt & Arreckx et al., Creation and analysis of biochemical constraint-based models: the COBRA Toolbox v3.0, Nature Protocols (accepted), 2018.

Cobrapy: A Solid Foundation for a Diverse Python Metabolic Modeling Community

Authors:

Beber, M. E., Technical University of Denmark
Cardoso, J. G. R., Technical University of Denmark
Diener, C., National Institute for Genomic Medicine
Fodor, M., Genialis, Inc
Kaafarani, A., Technical University of Denmark
King, Z. A., University of California, San Diego
König, M., Institute of Theoretical Biology, Humboldt University Berlin
Kutuzova, S., Technical University of Denmark
Lieven, C., Technical University of Denmark
Lopez, A., Technical University of Denmark
Mundy, M., Mayo Clinic
Rai, V., Department of Computational Medicine & Bioinformatics, University of Michigan
Redestig, H., Genencor International / DuPont
Sonnenschein, N., Technical University of Denmark
St. John, P. C., National Renewable Energy Laboratory
Steinbiß, S., Debian Project

When the COBRApy project began, its goals were defined as providing an intuitive, object-oriented interface for constraint-based metabolic modelling in Python that 1) can represent complex biological processes, 2) meets the computational challenges of next generation stoichiometric constraint-based models and omics data sets, 3) promotes related research

via freely available software, and 4) provides an interoperable alternative to the COBRA toolbox for Matlab.
Today, we can safely say that COBRApy has achieved all of those goals and has become an established tool that both academic and industry users rely on. We want to raise awareness for the maturity of COBRApy and how it has enabled productive constraint-based modeling by publishing a stable 1.0 release. In this work, we not only present the many improvements that have happened since the start of the project, but also highlight the vibrant and diverse community of Python packages built on top of COBRApy that together provide a plethora of methods related to constraint-based reconstructions and modeling.

While navigating a diverse software landscape akin to Bioconductor or the Python package index can be daunting at first, the benefits of distributed development are more publicly available tools due to the creative freedom provided, fast access to implementations of latest methods, and more direct contact between users and authors. Last but not least, a healthy open source community that encourages contributions from anyone can help mitigate the problem of abandoned academic software. Over the course of its six years of existence, COBRApy has been under the stewardship of several people and has seen contributions from almost 30 authors without whom it would not be where it is today. Here, we aim to provide a solid overview of the packages and capabilities within the community built around COBRApy.

Comparative Genomics and Network Modeling of Parasites

Authors:

Carey, M. A., University of Virginia
Papin, J. A., University of Virginia
Guler, J. L., University of Virginia
Untariou, A., University of Virginia
Stolarczyk, M., University of Virginia
Medlock, G. L., University of Virginia

Eukaryotic parasites cause diseases with large global burdens, resulting in over one million deaths annually. To develop novel antiparasitic drugs, experimental model systems often use alternative parasite species that are easier to manipulate genetically or to observe *in vivo* in mice, extrapolating these observations to disease-causing parasites. However, characterization of functional differences between parasite species is limited to *post hoc* and single-target studies, limiting the utility of these experimental models of disease and historic data. Each parasite genome encodes unique enzyme annotations; however, it is unclear whether these differences arise from divergent functions or incomplete genome annotation.

To address this challenge, we generated metabolic reconstructions from 160 parasite genomes, representing 38 genera and 111 species. Unsurprisingly, one of the largest genomes (*Chromera velia*) has the most unique reactions (35); however, the smallest genome (*Plasmodium billcollinsi*) has six unique reactions. Reconstruction and genome sizes are correlated, but even small networks contain unique features. We next applied our novel automated curation approach to leverage manual curation to improve reconstructions for related organisms. We used these semi-curated reconstructions to compare metabolic capacity and pathway utilization and generate hypotheses about species-specific functions. Here, we focus on the most lethal malaria parasite,*Plasmodium falciparum*, and the mouse model of severe malaria, *P. berghei*. For example, while *P. falciparum* grows in anucleated host erythrocytes, *berghei* prefers nucleated reticulocytes; *in silico, P. falciparum*has a reduced demand for host purines when compared to *P. berghei*, perhaps explaining this host cell preference and highlighting a pathway for which *P. berghei* is a poor model organism for*P. falciparum* antimalarial development. By performing these analyses with all 160 reconstructions, we developed a framework to identify metabolic discrepancies and commonalities between genera and species, facilitating comparison of experimental findings and optimizing model system selection to develop therapeutics for parasitic diseases.

Computation-Driven Mechanistic Understanding of the Cellular Cost and Regulation of Defensive Pigment Production

Authors:

Saha, R., University of Nebraska-Lincoln
Schroeder, W., University of Nebraska - Lincoln
Kumar, J., University of Nebraska - Lincoln
Harris, S., University of Nebraska - Lincoln

The endoplasmic reticulum (ER) is key to synthesizing mature, functional proteins. Its complex structure consisting of large thin sheets densely packed around the nucleus and elongated tubules reaching throughout the cell is key to its function, and is dynamically regulated in response to cellular signals. CLIMP-63 is the only known protein to play a role in shaping the ER, while also being the target of palmitoylation, one of the few reversible post-translational lipid modifications. In addition to its role in ER structure regulation, CLIMP-63 plays a role as a receptor at the cell surface, where it is known to bind to several ligands. These multiple roles make untangling the regulation of CLIMP-63 abundance and localisation a non-trivial task. We employed rule-based modelling to build a dynamical representation of CLIMP-63 birth, dimerisation, palmitoylation, traficking and degradation. Experimental data in the form of time-dependent radioactive labelling was used to calibrate the model with heuristic optimisation methods. Multiple predictions and hypothesis were generated as to the regulation of CLIMP-63 levels and localisation.

Sensitivity Analysis was used on the calibrated model to maximise the extraction of useful knowledge and hypotheses.

Computational Modeling of Ostrecoccus Tauri Central Metabolism Based on Statistical Thermodynamics

Authors:

Kumar, N., Pacific Northwest National Laboratory
Cannon, W. R., Pacific Northwest National Laboratory
Evans, J. E., Pacific Northwest National Lab
Zucker, J. D., Pacific Northwest National Laboratory

Developing predictive metabolic model based on the fundamental principles of physics is very critical to understand how cell structure affects function and subsequently investigate cellular dynamics relevant for bioenergy application. New computational tools are needed for quantitative modeling of cell that predict metabolite levels in high throughput manner, characterize thermodynamics and kinetics of individual reactions and energetics requirement of most likely metabolic pathways. In this talk, I will discuss our newly developed ODE-based optimization approach based on statistical thermodynamics for developing a central metabolic model of*Ostreococcus tauri,* the smallest known photoautotrophic eukaryote. We used the maximum entropy production principle to derive fluxes through a central metabolic network of *Ostreococcus tauri* . The predicted metabolite concentrations produced from maximum entropy production rate solution are compared to those typically expected from experiment using a loss function from which post-translational regulation of enzymes is inferred. Subsequently, we re-optimize the system with the inferred regulation and determine optimal rate constants for the metabolic network from the metabolite concentrations and reaction fluxes. At the end, I will discuss biophysical insights obtained from the modeling of *Ostreococcus tauri* that are relevant for biofuels production.

Computational Pathway Design for Funneling Lignin Intermediates to Aromatic Products

Authors:

Wang, L., The Pennsylvania State University
Beckham, G. T., National Renewable Energy Laboratory
Maranas, C. D., The Pennsylvania State University

The heterogeneity of the aromatic products from lignin catalytic depolymerization is one of the major challenges for lignin valorization. Microbes have evolved many catabolic pathways to tackle the challenge by funneling such heterogeneous intermediates to a few central aromatic products, which lead further to intra- or extradiol ring opening to produce value-added chemicals. However, such funneling pathways are only well-characterized in a few organisms such as*Sphingobium* sp. SYK-6 and *Pseudomonas putida* KT2440. Further effort to explore novel bacterial funneling pathways is necessary. Herein, we apply the *de novo*pathway design tool (novoStoic [1]) to computationally prospect all possible pathways for lignin-derived mono- and biaryls. First, we augment the MetRxn database with a new dataset of elementally balanced reaction operators using the automated atom mapping based reaction rule extraction procedure termed rePrime. For each reaction, the rePrime procedure identifies and captures as reaction rules the molecular graph topological changes underpinning the substrate to product graph conversion. A reaction rule captures the location of active reaction centers affected by the conversion of substrates to products. The reaction rules and known reactions are then operated upon a mixed integer linear programming algorithm (novoStoic) to identify a mass-balanced biochemical network that converts a source to a target metabolite while minimizing the number of*de novo* steps. We demonstrate the application of novoStoic in a few case studies such as designing shorter pathways from ferulate to protocatechuate and exploring Cð□›¼-Cð□›½ cleavage pathway of guaiacylglycerol-ð□›½-guaiacyl ether. By exploring the uncharted chemical space enabled by enzyme promiscuity, novoStoic paves the way to identify more carbon/energy efficient lignin funneling pathways with minimal heterologous enzymes to engineer, while exploring the organismâ€™s potential underground metabolism.

References

[1] Kumar, A., et al. Pathway design using de novo steps through uncharted biochemical spaces. *Nature communications* 9.1 (2018): 184.

Constraint-Based Modeling of Allelic Variation Mechanistically Links Genome-Wide Associations

Authors:

Kavvas, E., UCSD

The advent of genome-sequencing technologies has provided datasets describing the genetic and phenotypic diversity of microbial species at an unprecedented scale. Statistical approaches have led the way in addressing this microbial data deluge, but lack the ability to reason mechanistically about genotype-phenotype associations. However, with the growing availability and scope of genome-scale reconstructions of microorganisms, a mechanistic analysis of these datasets may be possible. Here we realize such an analysis through development and application of an allele-centric constraint-based modeling framework. The framework is applied to a dataset of 1,595 drug-tested *M. tuberculosis* strains using both reconstruction and inference-based approaches. Known features of *M. tuberculosis* AMR evolution are recapitulated and new insights revealed.

Constraint-Based Modeling Reveals the Distinct Metabolic Potential in Gut Microbiomes of Inflammatory Bowel Disease Patients with Dysbiosis

Authors:

Heinken, A., University of Luxembourg
Heirendt, L., University of Luxembourg
Baldini, F., University of Luxembourg
Thiele, I., University of Luxembourg
Fleming, R. M. T., Leiden University

The human gut microbiome plays an important role in human health. It is well known that inflammatory bowel diseases (IBD) are associated with microbial dysbiosis and altered fecal metabolomics profiles, however, the etiology of this complex, multifactorial disease remains poorly understood.

Recently, we published AGORA, a resource of curated genome-scale metabolic reconstructions for 773 common gut microbial strains. Moreover, we have published a COBRA Toolbox extension that enables the creation of personalized microbiome models from metagenomics data, as well as a high-performance implementation of constraint-based modeling in Julia. Here, we systematically profile the metabolic capabilities of gut microbiomes in health and disease states.

We retrieved strain-level relative abundances from metagenomics data from a cohort of pediatric Crohnâ€™s Disease patients with and without dysbiosis and healthy control children (108 samples in total) and constructed a personalized microbiome model for each sample using the corresponding AGORA reconstructions. Subsequently, we predicted for each microbiome model the quantitative biosynthesis potential for all secreted metabolites as well as the strain-level contributions to each metabolite in each individual microbiome.

In total, 182 metabolites from diverse subsystems could be secreted by at least one individual microbiome, of which 124 had statistically significantly different production potential in dysbiotic IBD patient microbiomes. The production potential for potentially detrimental metabolites such as lactate, sulfide, trimethylamine, and acetaldehyde was increased in dysbiotic IBD microbiomes. These metabolites were mainly produced by Proteobacteria representatives in this group but not in non-dysbiotic IBD or healthy microbiomes. In contrast, dysbiotic IBD microbiomes were depleted in production potential of butyrate, and B vitamins due to lower abundances of the contributing taxa.

In summary, we present an efficient computational approach to systematically interrogate individual microbiome models. This approach enables us to mechanistically link disease-relevant metabolites with gut microbial taxa known to play a role in IBD.

Constructing a High Quality Genome-Scale Metabolic Network for Streptomyces Lividans TK24

Authors:

De Winter, K., KU Leuven, University of Leuven
Ghesquière, J., KU Leuven, University of Leuven
Simoens, K., KU Leuven, University of Leuven
Busche, T., University of Bielefeld
Rückert, C., University of Bielefeld
Shlomi, T., Technion IIT
Kalinowski, J., University of Bielefeld
Bernaerts, K., KU Leuven, University of Leuven
Daniels, W., KU Leuven, University of Leuven
Lievens, B., Laboratory for Process Microbial Ecology and Bioinspirational Management,Department of Microbial and Molecular Systems

Streptomyces lividans is increasingly attracting attention as a potential industrial host for heterologous production of proteins and secondary metabolites. The current interest in *S. lividans*, combined with the increasing relevance of constraint-based modeling for strain improvement, urges for the creation of a high-quality genome-scale metabolic network (GSMN).

The most recent GSMN of model organism *S. coelicolor*â€" with high genomic similarity to *S. lividans*â€"is taken as a starting point for model construction. Comparative genomics is used for the conversion into *S. lividans*gene-reaction relationships. Reactions which lack *S.*

lividans enzymes are removed when this did not result in dead-end reactions with expressed (RNA-seq) enzymes or loss of growth on known growth substrates did not occur. Furthermore, KEGG reaction identifiers are added to the model, and KEGG metabolite identifiers and Enzyme Commission numbers are re-evaluated.

The model is extended using genomic, metabolomic and phenomic data. From the genome, metabolic reactions catalyzed by predicted enzymes not yet included in the model are added, as well as their substrates if they are not yet included as metabolites. Furthermore, additional metabolites, detected through LC-MS, are linked to the model by addition of metabolic reactions whenever this is reasonable with respect to the in silico phenotype and literature. Biolog Phenotype MicroArrays are used for testing respiration of *S. lividans* TK24 on 348 substrates and subsequent gap-filling of the model. To maintain high quality, the library for gap-filling is limited to 13 selected published bacterial GSMNs. Gap-filling was performed by searching for minimal sets of reactions that restore growth on a given substrate through mixed integer linear programming, and selecting the most appropriate solution. [Research funded by EU FP7-KBBE-2013-7 StrepSynth (grant n°613877).]

Construction of a Dynamic Flux Balance Analysis Model for Streptococcus Gordonii

Authors:

De Winter, K., KU Leuven, University of Leuven
Bernaerts, K., KU Leuven, University of Leuven
Waldherr, S., KU Leuven
Teughels, W., KU Leuven

Periodontitis is a chronic and inflammatory disease of the tooth supporting tissues. Environmental changes can bring oral biofilms from a healthy state to dysbiosis, which potentially evolves to periodontitis. Overall, we aim to construct a biofilm community model that brings more insight in dysbiosis of the oral biofilm and the influence of environmental factors. We develop a dynamic flux balance analysis (DFBA) model for *Streptococcus gordonii* which forms the first step in the development of an oral biofilm community model linked to periodontitis.

S. gordonii produces H_2O_2 and lactate, inhibiting pathogen outgrowth and enabling cross-feeding in the biofilm, respectively. First, the automated reconstruction taken from the AGORA database[1] was curated using literature information and databases (e.g., KEGG, RHEA, BIGG). FBA was used to evaluate growth rates, and to analyze carbon routing through the central carbon metabolism and amino acid biosynthesis pathways, resulting in a GSMN with 1278 reactions and 1078 metabolites, reproducing realistic FBA predictions. Secondly, the GSMN was transformed into a planktonic model in DFBAlab[2]. Simulations with glucose as main C-source predict the expected coupled production of lactate and H_2O_2, and show that protoheme depletion diverts carbon resources to lactate, thus triggering lactate production. Protoheme is an essential compound that cannot be synthesized by *S. gordonii*. H_2O_2 production is predicted with and without protoheme limitation, although concentrations are low. Analyses of growth on amino acids showed enhanced production of $H_2O_2$for growth on glucose and arginine, without influencing lactate production. Finally, DFBA simulations in a *S. gordonii* biofilm are established. Simulations still show the coupled production of lactate and H_2O_2. In a following stage, additional community members are added to the model.

[1]S. Magnúsdóttir et al., Nat. Biotechnol., vol. 35, no. 1, 2017.
[2]J. A. Gomez, K. Höffner, and P. I. Barton, BMC Bioinformatics, vol. 15, no. 1, 2014.

Data-Driven Discovery of Novel Therapeutic Targets Against Community-Associated Methicillin-Resistant Staphylococcus Aureus

Authors:

Islam, M. M., University of Nebraska-Lincoln
Saha, R., University of Nebraska-Lincoln
Thomas, V. C., University of Nebraska Medical Center

We recently uncovered that cellular metabolism is geared towards growth maximization but constrained by an upper rate of cellular Gibbs energy dissipation. Exploiting this principle in flux balance analysis generated excellent predictions of exchange and intracellular fluxes across a wide range of conditions and carbon sources, and even including some metabolite concentrations. This method requires a thermodynamically consistent formulation of cellular processes including Gibbs energies of reaction for every metabolic process, and a Gibbs energy balance, which states that the Gibbs energy dissipated by all cellular processes is equal to the Gibbs energy exchanged with the environment. We here present a detailed workflow, exemplified for a genome-scale metabolic model of *Escherichia coli,* how to implement and apply this method starting from any stoichiometric metabolic reconstruction. This workflow encompasses the formulation of such a model, the parameterization using experimental data together with regression analysis and the application in flux balance analysis. Furthermore, we present strategies how to solve the required nonlinear and nonconvex optimizations. Given the limited amount of required input data, and the precision and extent of the model predictions, we consider this method a valuable addition to current flux balance analysis approaches.

Deconvoluting Independent Regulatory Signals in the Escherichia coli Transcriptome

Authors:

Sastry, A., University of California, San Diego
Palsson, B. O., University of California, San Diego
Gao, Y., UC San Diego
Szubin, R., University of California, San Diego
Hefner, Y., University of California, San Diego
Xu, S., University of California, San Diego
Kim, D., University of California, San Diego
Choudhary, K. S., University of California, San Diego
Yang, L., University of California, San Diego
King, Z. A., University of California, San Diego

Underlying cellular responses is a transcriptional regulatory network (TRN) that modulates gene expression. Current descriptions of the TRN cannot quantitatively deconvolute the transcriptome into the relative contributions of individual transcriptional regulators. Here, we applied unsupervised learning to a compendium of high-quality *Escherichia coli* RNA-seq datasets (median $R^2 = 0.99$ between biological replicates) to identify 71 statistically independent regulatory signals. Summation of the 71 signals explained over 80% of expression variation across 115 unique experimental conditions. Of these, 50 signals were directly linked to characterized transcriptional regulators. Condition-specific signal strengths were validated by exposure to new environmental conditions, confirming 76% of predicted signal activations. The resulting decomposition of the transcriptome provided: 1) a quantitative, mechanistic explanation of responses to environmental and genetic perturbations, 2) a guide to gene function discovery, 3) characterization of a novel pyruvate-responsive transcription factor, and 4) a basis for comparing transcriptional regulation across closely related strains. Thus, we find that signal summation forms an underlying principle that can describe the composition of the prokaryotic transcriptome.

Describing the Glucose-Lactate Consumption Rate during Expansion and Osteogenic Differentiation of Human Bone Marrow Derived Mesenchymal Stem Cells: A Premise for Building a Systems Biology Model of Osteogenesis Using Metabolomics Analysis

Authors:

Sigmarsdóttir, Þ. B., Reykjavík University
Sigurjonsson, O., Reykjavik University
McGarrity, S., University of Iceland
Rolfsson, Ó., University Of Iceland

Transplant of cells derived from mesenchymal stem cells to treat diseases such as osteoporosis is tantalisingly close. The production of cells for transplant requires further study and currently there is lack of knowledge on metabolomic changes that occur during expansion and osteogenic differentiation (OD) of MSCs. It has been shown that mesenchymal stem cell osteogenic differentiation is accompanied by metabolic shifts, especially the utilisation of glucose. In order to better understand this process and expand the knowledge of other metabolic pathways that interact with stem cell fate this project is building constraint based metabolic models to integrate transcriptomic and metabolomics data.

Publically available transcriptomic data and new glucose and lactate measurements were combined in the CobraToolbox. This created initial models of the first stages of expansion and osteogenic differentiation based. Functional assays confirmed the process of osteogenesis *in vitro*. Glucose and lactate measurements from six donors during expansion and osteogenic differentiation indicate multiple stages to these processes. . The final stage of expansion shows an increase in the use of glucose, without an increase in lactate production. In osteogenesis the average glucose consumption and lactate production measures higher over the later 14 days, giving an indication of the possibility of two phenotypes.

The models were able to recreate known metabolic features of mesenchymal stem cells during expansion and osteogenesis. These include relative levels of oxidative phosphorylation to glycolysis and production of kynurenine from tryptophan. These models also suggest glycan production as well as more central metabolism as areas key to the shift from expansion to osteogenic differentiation.

Genome scale metabolic modelling offers a way of combining multiple types of omics data in order to better understand and manipulate the features of a biological system.

Determination of CHO Biomass Composition

Authors:

Szeliova, D., University of Natural Resources and Life Sciences
Ruckerbauer, D., The Austrian Centre of Industrial Biotechnology
Hanscho, M., The Austrian Centre of Industrial Biotechnology
Galleguillos, S., University of Natural Resources and Life Sciences
Borth, N., University of Natural Resources and Life Sciences
Hann, S., University of Natural Resources and Life Sciences

Chinese hamster ovary (CHO) cells are the primary host organism for the production of protein biopharmaceuticals. Significant improvements in product yield and cell growth were achieved in the past years by bioprocess and media optimization, directed evolution and targeted genetic engineering. However, a deeper understanding of the underlying processes in the cells is still limited. Recently, a CHO-specific genome scale metabolic model was created in a large community effort. This model is a comprehensive resource of CHO metabolism. Using COBRA methods, we are now starting to get valuable insights into the cellsâ€™ metabolism, their protein production capabilities and potential limitations. One essential input for the model is biomass composition. It has been shown that using strain and condition specific biomass together with bioprocess data improves the modelâ€™s predictions. Currently, however, the model uses estimates and literature values, since comprehensive data about CHO cell composition, specifically of individual cell lines or strains, are lacking. In this work, methods for the determination of CHO biomass composition (proteins and amino acids, lipids, DNA, RNA and cell dry mass) were established. These include chromatography and mass spectrometry determination of amino acids, fluorimetric and spectrophotometric quantification of nucleic acids and gravimetric quantification of cell dry mass. For the first time, biomass compositions of various CHO host and producer cell lines in exponential phase were accurately characterized in media with or without glutamine. The goal was to assess variability of the biomass composition among different strains and conditions to see which components are the most variable and therefore should be measured every time and which could be generalized for all CHO cells.

Determination of Growth-Coupling Strategies and Their Underlying Principles

Authors:

Ebert, B. E., RWTH Aachen University
Alter, T. B., RWTH Aachen University
Blank, L. M., RWTH Aachen University

Many insects of medical and agricultural importance depend on symbiotic associations with nutrient producing bacteria. Very commonly, net nutrient production by the bacteria is dictated by metabolic competition and cooperation among multiple bacterial taxa and strongly influenced by the metabolites derived from the insect host, but the mechanistic details of the multi-way metabolic interactions are poorly understood. To investigate the processes shaping insect-microbe metabolic interactions and to derive estimates of the composition and amount of nutrients exchanged between insect and microbe, we reconstructed genome and transcriptome-informed metabolic models of sap-feeding insects that harbor intracellular bacterial partners and the bacterial community in the Drosophila gut. Our simulations reveal that among intracellular bacterial partners the insect host controls bacterial production of nutrients, by precise controls over the concentrations of substrate metabolites in essential nutrient biosynthetic pathways. The net result is the structuring of the metabolic networks of the bacteria, to promote cooperative cross-feeding of metabolites and minimize competition for host-derived substrates. Among gut microbes competitive interactions dominate and the quality and quantity of nutrients available to the host is influenced by the composition of gut microbiota which synthesize and consume host and microbe derived nutrients. Our studies elucidate the metabolic basis of insect-microbe interactions and provide the methodology to identify specific gene targets for novel pest control strategies and to select the optimal microbial consortia for personalized microbial therapies in treating animal diseases associated with microbial dysbiosis.

Developement and Application of Constraint-Based Modeling Methods to Study Vulnerabilities Associated to Lipid Metabolism in Prostate Cancer

Authors:

Marín de Mas, I., Technical University of Denmark

Aguilar, E., University of Barcelona

Bedia, C., CSIC

Thomson, T. M., Molecular Biology Institute of Barcelona, National Research Council (IBMB-CSIC)

Tauler, R., CSIC

Papp, B., Biological Research Centre of the Hungarian Academy of Sciences

Cascante, M., University of Barcelona

Nielsen, L. K., Technical University of Denmark

It is increasingly clear that significant alterations in the lipid profile of cancer cells accompany tumor progression and metastasis. These changes are induced by a metabolic reprogramming that enhances the malignant phenotype of cancer cells. In this context, genome-scale metabolic models (GEM) have emerged as a valuable platform to integrate different omic data to study cancer metabolism from a holistic perspective.

Here we present a constraint-based model-driven approach integrating transcriptomic data to study the metabolic profile of two clonal sub-populations from a prostate cancer cell line (PC-3): PC-3/M and PC-3/S, representing pre and post Epithelial-Mesenchimal-Transition stages. This model-driven analysis and experimental validations unveiled a marked metabolic reprogramming in long-chain fatty acids metabolism.

While PC-3/M cells showed an enhanced entry of long-chain fatty acids into the mitochondria, PC-3/S cells used this pool as precursors of eicosanoid metabolism. This metabolic reprogramming endows PC-3/M cells with augmented energy metabolism for fast proliferation and PC-3/S cells with increased eicosanoid production impacting angiogenesis, cell adhesion and invasion.

These findings highlight the relevance of lipid metabolism in cancer development and progression. However, lipid-associated pathways are poorly annotated in human GEMs which limits the use of this tool. To overcome this limitation, we developed an algorithm-based metabolic network building method to enrich existing GEMs with lipid-associated pathways. This algorithm was applied in the study of the metabolic profile of another prostate cancer cell line (DU145) in response to a chronic exposure to different endocrine disruptors inducing malignancy and alterations in the lipid profile. The resulting lipid-enriched GEM covered 98% of the altered lipids compared to only 5% in the original model. Thus, this approach improves lipidomic data integration into GEM reconstructions, enabling a more in-depth study of the mechanisms underlying diseases with a strong metabolic component such as cancer.

Developing and Evaluating Integrated Metabolic Regulatory Models for Microbial Life

Authors:

Tintle, N., Dordt College
DeJongh, M., Hope College
Lindsey, W., Dordt College
Best, A., Hope College
Creagar, M., University of San Francisco
Disselkoen, C., University of California, San Diego
Fore, R., Brown University
Friend, D., University of Nevada, Reno
Kamp, T., Dordt College
Henry, C. S., Argonne National Laboratory

A number of papers have described methods to include transcriptional regulatory networks (TRNs) in the development of metabolic models. However, in general, these models do not allow for statistical uncertainty in TRN interactions. Furthermore, methods which do allow for uncertainty do so in a non-rigorous manner which, effectively, downweights prior transcriptomics data to the point where it has little impact on the resulting model. To combat these limitations, we developed an integrated metabolic regulatory model (iMRM). This novel approach explicitly models the statistical uncertainty in gene state activity inferences by using gene activity state estimates, and the iMRM process measures the flux through reactions for a sole single-cell organism. This allows for benefits such as differential thresholds for gene activity for different genes, a stronger correlation with experimental flux data, and the application of a non-Boolean confidence metric. Furthermore, this is the first such metabolic model to be grounded on a statistically rigorous foundation.

Development of an Accelerated Workflow for Parameterizing Kinetic Models of Metabolism

Authors:

Gopalakrishnan, S., The Pennsylvania State University
Dash, S., The Pennsylvania State University
Maranas, C. D., The Pennsylvania State University

Kinetic models can simulate an organismâ€™s metabolism under genetic perturbations by capturing the changes in metabolite pools and the impact on reaction fluxes in the form of regulations. The challenge of estimating kinetic parameters can be resolved using the Ensemble modeling (EM) approach which samples parameters consistent with a training dataset. However, the EM approach implemented using GA is computationally expensive. Thus, we have described here a gradient-based kinetic parameterization workflow that exploits the strengths and overcomes the shortcomings of the EM approach. First, the kinetic parameters are anchored to the steady-state metabolic flux distribution in the wild-type strain by exploiting the rate laws governed by mass-action kinetics relating fluxes through elementary reactions to sampled enzyme levels and kinetic parameters. Following this, a genetic perturbation is introduced by adjusting the total abundance of specific enzymes relative to the wild-type strain so as to reflect the *in vivo* genetic perturbation in the form of up/down-regulations or gene deletions. The steady-state fluxes in the mutant conditions are evaluated by decomposing the underlying system of bilinear equations into two linear sub-problems and iterating between them until convergence is achieved. Based on the deviation of predicted fluxes from measured data, kinetic parameters are updated using a Newtonian step until optimality is achieved. This accelerated procedure has been applied to a *C. thermocellum* core model (106 reactions, 72 metabolites, 674 elementary kinetic parameters, and 4 mutant conditions) and an *E. coli* near genome-scale model (319 reactions, 276 metabolites, 2536 elementary kinetic parameters and 7 mutant conditions). Parameterization was completed in 12 hours for *C. thermocellum* and 48 hours for *E. coli*, compared to a computational time requirement of 1 week and 6 weeks for *C. thermocellum* and *E. coli*, respectively using the GA-based procedure.

Dynamic FBA with Time-Course Transcriptomics

Authors:

Sulheim, S., SINTEF Industry
Almaas, E., NTNU - Norwegian University of Science and Technology
Wentzel, A., SINTEF Industry

Genome-scale metabolic models provide a framework for the computational prediction of a wide range of phenotypes, such as growth in different nutrient environments, and effects of genetic manipulations, as well as aiding in the interpretation of â€™omics data. Dynamic flux balance analysis (dFBA) is a method where the dynamics of growth is incorporated into the framework of steady-state flux balance analysis (FBA) by updating environmental conditions at regular time-intervals. Here, we present tt-dFBA by extending the dFBA methodology through incorporating time-course transcriptome data to account for changes in gene expression levels during batch fermentation.
We apply our framework to the case of batch fermentation of Streptomyces coelicolor, using the recent reconstruction iKS1317, to explain the observed metabolic switching and onset of secondary metabolite production when the organism faces phosphate depletion. Our results provide insights into the metabolic flux rerouting causing the observed metabolic changes.

Effective Visualization for Investigating Elementary Flux Modes in Genome-Scale Metabolic Models

Authors:

Sarathy, C., Maastricht University
Lenz, M., Maastricht University
Kutmon, M., Maastricht University
Evelo, C. T., Maastricht University
Arts, I. C. W., Maastricht University

Elementary Flux Modes (EFMs) are an indispensable tool for constraint-based modeling and metabolic network analysis[1]. EFMs allow exploration of cellular metabolism beyond conventional pathway definitions and integration with high-throughput data. Since the field has been driving towards improving algorithms to enumerate all (or subsets) of EFMs, their visualization has been limited to mapping only the flux data in standalone tools. However, flux mapping is just one component of systems studies, where several molecular datatypes are combined. There is a need for an efficient visual analysis approach that complements current efforts integrating omics data with genome-scale metabolic models. In this study, we developed a simple, efficient and MATLAB-based workflow for graphically visualizing EFMs as a network of reactions, metabolites and genes. Our flexible workflow seamlessly integrates COBRA[2] or RAVEN[3] with the open-source tool, Cytoscape[4]. Cytoscape not only enables data visualization and advanced network analysis, but also network extension with other molecular interaction datatypes, all within a single framework. Once processed through our workflow, SBGN layout is automatically applied on the visualized network requiring little-to-no manual effort by the user. Furthermore, combining existing tools avoids additional installation of flux-mapping tools, used previously, thus making it both user-friendly and time-saving. We illustrate our workflow using a subset of EFMs generated from both small and large models and map gene expression data on the visualized EFMs, thereby, demonstrating that efficient, integrative visualization can allow for a better understanding of biological processes.

References: [1] Klamt, S. *et al.*,DOI:10.1371/journal.pcbi.1005409.[2] Schellenberger J. *et al.*,DOI:10.1038/nprot.2011.308. [3] Agren, R. *et al.*,DOI:10.1371/journal.pcbi.1002980. [4] Shannon, P. *et al.*,DOI:10.1101/gr.1239303.

Elementary Mass Action Stoichiometric Simulation Models Predict Non-Negligible Concentrations of Enzyme-Bound Metabolites

Authors:

Matos, M. R. A., Technical University of Denmark
Zielinski, D. C., University of California
Nielsen, L. K., Technical University of Denmark
Sonnenschein, N., Technical University of Denmark

Improvement of cell factory performance, to the point where production of chosen molecules becomes economically profitable, requires extensive metabolic modification. Metabolic models have been used to both provide new insights into the inner workings of metabolism and new directions for strain engineering. Kinetic models in particular are key to model the dynamics of metabolism and substrate-level enzyme regulation.

Here, we use the elementary Mass Action Stoichiometric Simulation (eMASS) framework to build a prototype kinetic model ensemble for glycolysis in *E. coli*. Following eMASS, we first build an individual kinetic model for each reaction by decomposing it into elementary reactions, according to the respective reaction mechanism. Next, the associated elementary rate constants are fitted to kinetic data obtained from the literature, e.g. k_{cat}, K_m, K_i, while satisfying the respective Haldane relationship. Finally, we assemble the systemâ€™s kinetic model by combining the individual reaction models and integrating fluxomics and metabolomics data. Since we model both free and enzyme-bound

metabolite concentrations explicitly, we drop the common assumption that enzyme-bound metabolite concentrations are negligible (x_{tot} â‰ˆ x_{free}) and predict the free metabolite concentrations.
Our preliminary results show that the percentage of enzyme-bound metabolite vs. total metabolite can be as high as 40%, i.e. the amount of enzyme-bound metabolite may not be negligible even in glycolysis. These results derive from the fact that in some cases the concentration of metabolites is in the same order of magnitude as the enzymes with which they interact.

Elucidating the Metabolic Rewiring of Pluripotent Stem Cells during Differentiation into Adult Progenitors Using an Updated and Enzyme-Constrained Human Genome-Scale Model

Authors:

Lakshmanan, M., Bioprocessing Technology Institute
Lee, D. Y., Bioprocessing Technology institute, A*STAR
Koduru, L., National university of Singapore
Pluripotent stem cells (PSCs) offer new potentials for treating multiple diseases via gene therapy and regenerative medicine due to their unique regenerative abilities. PSCs differentiate into many distinct germ layer cells and produce various human organs via various transcriptional, epigenetic and metabolic changes. While the transcriptional and epigenetic changes during stem cell differentiation is widely researched, the metabolic changes of PSCs during differentiation remain largely uncovered, except the well-known switch from aerobic glycolysis into oxidative phosphorylation for energy generation. Therefore, we utilize a systematic approach combining transcriptomic data and the human genome-scale model (GEM) to examine the metabolic signatures of pluripotent stem cells, germ layer cells, and adult stem cells in comparison to various mature cells. Here, it should be highlighted that although the availability of several human GEMs has substantially

improved our understanding on the molecular basis of various diseases and phenotypes, we still observed the variability in internal metabolic fluxes predicted by the models, across conditions due to the limited constraints typically used in constraints-based flux analysis. To address this critical issue, here, we present the most comprehensive and biochemically consistent human GEM to-date, Recon3E, and included the necessary kinetome information to enable its ready use during the simulations, thereby improving the phenotype predictions further. Recon3E accounts for 3463 open reading frames, 7182 reactions corresponding to 987 unique enzymes, and 3023 unique metabolites. Importantly, Recon3E contains the kinetic information, i.e. k_{cat}, for 672 of the total 987 enzymes where 74% of these values are human-specific. Using Recon3E, we were able to predict the metabolic rewiring which occurs during the differentiation of PSCs into adult progenitors, emphasizing the role of unsaturated lipids, amino acid derivatives and one-carbon metabolites in maintaining self-renewal and pluripotency, in addition to the commonly cited glycolysis and oxidative phosphorylation.
(Funding: SSAC# PJ01334605, Republic of Korea)

Enzyme Multicollinearity in Genome-Scale Metabolic Models Revealed By an Efficient Coupling Algorithm

Authors:

Pradhan, D., University of Illinois at Urbana-Champaign
Jensen, P. A., University of Illinois at Urbana-Champaign

Numerous reactions control bacterial behavior and homeostasis. Dimensionality reduction is often necessary to analyze COBRA models, especially non-convex models that require intensive computations. One method to reduce networks is to identify reactions with colinear, or fully coupled, fluxes. In a COBRA model, coupled reactions can be identified using the Flux-Coupling Finder (FCF) algorithm. However, identification of coupled reaction sets is not as useful as identification of coupled enzyme sets, since enzymes are targets of metabolic engineering and drug discovery efforts. To calculate coupling between enzymes, we developed a mathematical transformation linking reaction fluxes and enzyme activity while preserving the nonlinear mapping between genes, proteins, and reactions. This framework, called Flux and Activity Linked Constraints (FALCON) expands the original linear program to a larger mixed-integer linear program (MILP). Although solving MILPs requires significantly more computational power, we developed an efficient variant of FCF (cached-FCF) that drastically reduces the runtime required to identify couplings in both COBRA and FALCON models.

We tested cached-FCF on the FALCON model for the bacterium *Pseudomonas aeruginosa.* We discovered that the enzymes in coupled sets showed significantly higher correlation of expression and fitness than sets of enzymes created from sets of coupled reactions. The coupled enzyme sets form the functional units of the organismâ€™s metabolic network. We further tested the effects of perturbing the metabolic network with single-gene knockouts. We observed that gene deletions force additional couplings among enzyme sets. These additional couplings link enzyme sets with similar expression or fitness patterns.

Erythrodb: A Multi-Omic Visual Knowledge Base for the Human Red Blood Cell

Authors:

Yurkovich, J. T., University of California, San Diego
Palsson, B. O., University of California, San Diego
As the life sciences have embraced the information age, we have witnessed the birth of many outstanding resources such as BioCyc, UniProt, the Protein Databank, and KEGG. With the continued generation of -omics data, we are faced with challenges in contextualizing, visualizing, and extracting information from these data. Here, we present ErythroDB (http://erythrodb.org), a multi-omic visual knowledge base for the biochemistry and metabolism of the human red blood cell with applications for transfusion medicine. This resource contains three primary forms of content: (1) an interactive metabolic map enhanced with data from the literature and information from external databases, (2) an interactive â€œbibliomeâ€□ that contains manually-curated literature on red blood cell metabolism and storage, and (3) a data repository that provides access to published -omics data sets. ErythroDB captures the true systems aspect of physiology by allowing multiple data sources to be visualized together. Moving forward, contextualizing findings with a systems-level view will help synthesize knowledge generated by the community, eventually leading to a better understanding of red blood cell physiology.

Essential Metabolism for a Minimal Cell - Poster

Authors:

Breuer, M., University of Illinois at Urbana-Champaign

The question of the core requirements of cellular life led to the construction of the "minimal cell" JCVI-syn3A: A cell where practically all genes were removed that were not essential for robust growth in a stress-free laboratory environment. With only 493 genes in a 543 kbp genome, JCVI-syn3A has a genome smaller than that of any independently-replicating cell found in nature. It provides a versatile platform to study the basics of cellular life and is small enough that a complete description of all cellular functions can be pursued. Here, we present an extensively curated metabolic reconstruction and flux balance analysis (FBA) model of this minimal cell, using the vast amount of experimental information available on its natural precursor, Mycoplasma mycoides capri. The model, featuring 337 reactions involving 304 metabolites, is near-complete with 98 % of enzymatic reactions justified through gene assignments and/or experimental evidence, and agrees well with gene essentiality data from transposon mutagenesis experiments. The 155 genes included in the reconstruction have a high in vivo essentiality or quasi-essentiality of 92 %, compared to 79 % in silico essentialityâ€"underscoring the minimality of the network. The reconstruction itself and the comparison of in vivo and in silico essentialities lead to new hypotheses on particular metabolic functions, suggesting specific experiments. Thus, the model provides a solid foundation for further experimental and computational studies on the minimal cell.

Etfl: A Formulation of ME-Models Accounting for Expression and Thermodynamics in Constraint-Based Models

Authors:

Salvy, P., EPFL
Hatzimanikatis, V., Swiss Federal Institute of Technology (EPFL)

Since flux balance analysis (FBA) has been introduced in Systems Biology, several attempts have been made at supplementing it with expression data [1-3]. With the notable exception of Oâ€™Brien et al. [1], none of them account directly for the enzyme production as a part of the mathematical programming formulation of the problem. Furthermore, Oâ€™Brien et al. propose a bilinear programming formulation that is computationally heavy.

We propose a top-down model formulation, from metabolism to RNA synthesis, that allows the simulation of thermodynamics-compliant intracellular fluxes, as well as enzyme and mRNA levels. The formulation results in a mixed integer linear problem (MILP) which enables both relative and absolute metabolite, proteins, and mRNA concentration integration. The proposed formulation does not require any bilinear solver. We present here the results obtained using the *Escherischia coli* model iJO1366 [4]. We show that the formulation is able to reproduce proteome-limited growth, which flux balance analysis cannot [1], as well as predict feasible mRNA and enzyme concentrations in the cell.

1. O'Brien, E.J., et al., *Genome-scale models of metabolism and gene expression extend and refine growth phenotype prediction.* Mol Syst Biol, 2013.**9**: p. 693.
2. SÃ¡nchez, B.J., et al., *Improving the phenotype predictions of a yeast genomeâ€□scale metabolic model by incorporating enzymatic constraints.* Molecular Systems Biology, 2017. **13**(8).
3. Beg, Q.K., et al., *Intracellular crowding defines the mode and sequence of substrate uptake by Escherichia coli and constrains its metabolic activity.* Proc Natl Acad Sci U S A, 2007. **104**(31): p. 12663-8.
4. Orth, J.D., et al., *A comprehensive genome-scale reconstruction of Escherichia coli metabolism-2011.* Molecular Systems Biology, 2011. **7**.

Exploring Biosynthetic Pathways for the Production of Five Methyl Ethyl Ketone Precursors

Authors:

Miskovic, L., EPFL

Tokic, M., Swiss Federal Institute of Technology (EPFL)

Hadadi, N., Swiss Institute of Bioinformatics (SIB)

Ataman, M., EPFL

Neves, D., RWTH Aachen University

Ebert, B. E., RWTH Aachen University

Blank, L. M., RWTH Aachen University

Hatzimanikatis, V., Swiss Federal Institute of Technology (EPFL)

Limited reserves of oil and natural gas and the establishment of new environmental policies sparked off intensive research towards sustainable production of the 2[nd]generation of biofuels, with Methyl Ethyl Ketone (MEK) being one promising fuel candidate. MEK is a commercially valuable petrochemical with an extensive application as a solvent. However, as of today, a sustainable and economically viable production of MEK has not yet been achieved despite several attempts of introducing biosynthetic pathways in industrial microorganisms. We used BNICE.ch as a retrobiosynthesis tool to discover all novel pathways around MEK. Out of 1â€™325 identified compounds connecting to MEK with one reaction step, we selected 3-oxopentanoate, but-3-en-2-one, but-1-en-2-olate, butylamine, and 2-hydroxy-2-methyl-butanenitrile for further study. We reconstructed 3â€™679â€™610 novel biosynthetic pathways towards these 5 compounds. We then embedded these pathways into the genome-scale model of *E. coli*, and a set of 18â€™622 were found to be most biologically feasible ones based on thermodynamics and their yields. For each novel reaction in the viable pathways, we proposed the most similar KEGG reactions, with their gene and protein sequences, as candidates for either a direct experimental implementation or as a basis for enzyme engineering. Through pathway similarity analysis we classified the pathways and identified the enzymes and precursors that were vital for the production of the target molecules. These retrobiosynthesis studies demonstrate the potential of BNICE.ch for discovery, systematic evaluation, and analysis of novel pathways in synthetic biology and metabolic engineering studies.

Exploring the Metabolic Role of Multidrug-Resistance Efflux Pumps in Salmonella Typhimurium

Authors:

Mack, S. G., University of Maryland
Dwyer, D. J., University of Maryland
Wang-Kan, X., University of Birmingham
Piddock, L. J. V., University of Birmingham

The surge in antibiotic resistance poses a serious threat to human health. In Gram-negative pathogens, diverse resistance-nodulation-division (RND) efflux pumps provide broad-spectrum resistance and therefore greatly contribute to antibiotic treatment failure **[1]**. While the structure and protective roles of these pumps has been established **[2]**, their homeostatic contributions are poorly understood. Recent work has suggested a broader metabolic role for the AcrB efflux pump in *Salmonella enterica* Typhimurium, with AcrB being the most common member of the RND family. The loss of AcrB efflux, but not expression, led to a loss in virulence and the downregulation of various pathogenicity factors **[3]**. Given that AcrB is highly conserved across all *Enterobacteriaceae* **[2]**, further exploration of its potential metabolic role is clearly needed.

To address this gap in understanding, we have developed a computational framework to predict and compare the metabolic behaviors of wild-type and the loss-of-function AcrB mutant of *S*. Typhimurium. Specifically, we analyzed RNAseq data from *S*. Typhimurium WT and mutant strains grown in minimal media for differentially expressed genes. Using a modified version of the metabolic transformation algorithm (MTA) **[5]**, transcriptomic data were integrated with a genome-scale metabolic model for *S*. Typhimurium **[6]** to generate metabolic flux predictions for each strain. From these flux profiles, we identified genetic perturbations that would shift WT metabolism closer to mutant metabolism (and vice versa). Our findings suggest the loss in AcrB efflux leads to increased fatty acid biosynthesis which may contribute to the overall loss in virulence. To validate these predictions, we are currently testing our top targets via metabolomics and infection assays.

1. Blair, J. M. A. et al. Nature Reviews Microbiology (2015).
2. Du, D. et al. Biological Chemistry (2015).
3. Wang-Kan, X. et al. mBio (2017).
4. Yizhak, K. et al. Nature Communications (2013).
5. Thiele, I. et al. BMC Systems Biology (2011).

Fine Tuning Thresholds to Facilitate Integration of Transcriptomics Data

Authors:

Joshi, C., University of California, San Diego
Schinn, S. M., University of California, San Diego
Chiang, A., University of California, San Diego
Richelle, A., University of California, San Diego
Lewis, N., University of California, San Diego

Integration of omics datatypes is becoming increasingly commonplace in the field of metabolic modeling. Among all of the omics datatypes, transcriptomics is the most chosen for integration with genome-scale metabolic models (GEMs) resulting in a wide array of algorithms. These algorithms are often preceded by preprocessing the transcriptomics data to generate list of high confidence genes or reactions. Multiple studies in the past have identified parameters related to preprocessing to be important, specifically thresholding of gene expression profile. However, it is not clear how thresholds should be decided. Here, in our analysis we look at tissue-specific transcriptomic data from HPA and GTex and provided novel insights on what factors could determine thresholds and what are the largest sources of false positive and false negative predictions. Our analysis also provides novel insights on gene expression patterns of tissue-specific and housekeeping genes. Further, we also suggest improvements, validate using proteomics data and metabolic tasks. We also applied our thresholding method to cancer cell-line transcriptomics data sets. Thus, we demonstrate better applicability of thresholds across different data sets.

Flux Balance Analysis and the Evolution of Xylose Fermentation in Yeasts

Authors:

Correia, K., University of Toronto
Khusnutdinova, A., University of Toronto
Li, P. Y., Merkle
Joo, J. C., Korea Research Institute of Chemical Technology
Brown, G., University of Toronto
Yakunin, A. F., University of Toronto
Mahadevan, R., University of Toronto

Growth kinetics follow lag, exponential, and stationary phases, with their transitions marked by drastic changes in biomass growth rates. This may be due to shifts in transcriptional activity, where high substrate concentrations promote increasing ribosome production to support the exponential phase. Under this hypothesis, there are smooth transitions from low to high growth rates on a phenotypically uniform population of cells. However, this is not the case for *Clostridium beijerinckii* which, when grown in batches, present various phenotypes coexisting. It starts as a vegetative phenotype, where fast substrate consumption is paired by fast biomass growth and production of acidic products. Acidification promotes a fraction of cells to murph into a clostridial phenotype. These are non-dividing cells whose metabolism is geared towards sporulation, a process where glucose and the previously released acidic species in the medium are consumed, producing in turn solvents. Released spores may later germinate into vegetative cells if conditions became favorable.

We propose a kinetic model where all three stages of *C beijerinckii* are present at any given time. The dynamics between phenotypic states are described by differential equations within dynamic FBA. To model the vegetative-clostrial and spore-vegetative transitions, we used the concentration of acidic species as arguments to a logistic equation whose value can be interpreted as the fraction of the vegetative cells transiting to clostridial state. Likewise, we used a second logistic equation, this time as a function of the concentration of solvents, to model the fraction of clostridial cells that change into spores.

Using public RNA-seq databases we determined divergently expressed genes in each phenotype. We pruned the metabolic network of *C beijerinckii*accordingly to create phenotype-specific metabolic networks. To train our model we measured changes in substrates, and acidic and solventogenic species along batch cultures of *C beijerinckii* under various conditions.

Gapseq: A Novel Approach for in silico Prediction and Analysis of Bacterial Metabolic Pathways and Genome-Scale Networks

Authors:

Waschina, S., Kiel University
Kaleta, C., Christian-Albrechts-University Kiel (UKSH Campus)
Zimmermann, J., Kiel University

In silico biochemical pathway analysis and metabolic phenotype prediction based on bacterial genome sequences became powerful research tools in biotechnology, medicine, and ecology. The accuracy of genome sequence-derived catalytic capability predictions and metabolic models are largely hindered by two factors: (i) False-positive and false-negative enzymatic reaction predictions due to incomplete knowledge of yet-unknown enzyme mechanisms, especially in non-model organisms, and (ii) inconsistencies in reaction databases, which frequently cause thermodynamically infeasible futile cycles and, hence, false phenotype predictions.

Here we present a new tool for pathway analysis and automated reconstruction of metabolic networks, called GapSeq, that addresses these obstacles. First, pathways are predicted by combining sequence homology search results to estimate the evidence of the presence of the pathwaysâ€™ individual reactions, pathway topologies, and information about the pathwaysâ€™ key reactions. Second, reactions predicted with high evidence scores are consolidated to construct draft genome-scale networks using a reaction database that has been manually curated and is completely futile cycle-free. This ensures, that also all possible reconstructions are free of thermodynamically infeasible substrate cycles. Third, a novel gap-filling algorithm was implemented that adds candidate reactions based on the reactionâ€™s evidence scores. Moreover, the gap-filling algorithm does not only enable biomass production under the defined media composition, but also fills gaps in alternative resource utilization pathways as well as anabolic pathways. This approach has the advantage that also gaps in the network are filled with candidate reactions that are only employed by the organism under specific environmental conditions or when interacting with different organisms.

We tested GapSeq on a large set of bacterial strains, which are physiologically well-described. We show that not only the presence of strain-specific pathways are correctly predicted, but also metabolic phenotypes, such as the production of specific fermentation products, are accurately predicted by GapSeq in combination with Flux Balance Analysis.

Genetic Optimization Algorithm for Metabolic Engineering Revisited

Authors:

Alter, T. B., RWTH Aachen University
Blank, L. M., RWTH Aachen University
Ebert, B. E., RWTH Aachen University

To date, several independent methods and algorithms exist exploiting constraint-based stoichiometric models to find metabolic engineering strategies that optimize microbial production performance. Optimization procedures based on metaheuristics facilitate a straightforward adaption and expansion of engineering objectives as well as fitness functions, while being particularly suited for solving problems of high complexity. Regarding the uprising of multi-scale models and a need for solving advanced engineering problems, we strive to advance genetic algorithms, which stand out due to their intuitive optimization principles and proven usefulness in this field of research. A drawback of genetic algorithms is that premature convergence to sub-optimal solutions easily occurs if the optimization parameters are not adapted to the specific problem. Therefore, we conducted comprehensive parameter sensitivity analyses focusing on the impact on finding optimal strain designs. We further demonstrate the capability of genetic algorithms to simultaneously handle (i) multiple, non-linear engineering objectives (ii) the identification of gene target sets according to logical gene-protein-reaction associations, (iii) minimization of the number of network perturbations, and (iv) the insertion of non-native reactions, while employing genome-scale metabolic models. This framework adds a level of sophistication in terms of strain design robustness, which is exemplarily tested on succinate overproduction in *Escherichia coli*.

Genome Scale Metabolic Analysis of the Lactobacillus Genus with a Focus on Probiotic Strains

Authors:

Moutinho, T. J. Jr., University of Virginia
Neubert, B. C., University of Virginia
Papin, J. A., University of Virginia

The Lactobacillus genus has, primarily, dominated the probiotic industry with a great deal of success in conferring health benefit as shown in a variety of rigorous clinical trials. In this study, we generate over 200 genome-scale metabolic reconstructions from a broad array of Lactobacillus species, including many of the currently sequenced species that are typically found in marketed probiotics, resident strains from human isolation sites, and fermented foods. We have conducted a strain level comparison of the reactomes for each genome, demonstrating that the metabolic reactome contributes to clustering of the genomes based on human health related isolation sites, including the gut, oral cavity, and vagina. Strains resident to the vaginal microbiota cluster separately from strains resident to the gastrointestinal tract, while there is a large overlap in metabolic potential between gut and oral residents. A random forest analysis was used to identify reactions specific to resident strains to the gastrointestinal tract when compared to other lactobacillus strains. Carbon and amino acid source utilization was predicted using the generated metabolic reconstructions, our data indicate that there is a great deal of variation among Lactobacillus strains with some correlation to isolation site. We present a phylogenetic analysis based on metabolic capabilities, demonstrating the similarities among Lactobacillus strains indicating which groups are most likely to compete in a microbial community. The metabolic reconstructions were generated using ModelSEEDs universal format, the PATRIC database, Mackinac, CobraPy, and Probannopy. This study provides a public resource of systematically generated genome scale metabolic reconstructions for a wide range of Lactobacillus strains for further curation and analysis of probiotics relevant to human oral, gut, and vaginal health. Additionally, this work lays the ground work for personalize design, selection, and delivery of Lactobacillus based probiotics.

Genome Scale Metabolic Model Assisted Strain Designs for Itaconic Acid Production in Yeast

Authors:

Zhao, Z., DSM Biotechnology Center

Young, E. M., Massachusetts Institute of Technology

Gielesen, B., DSM Biotechnology Center

Wu, L., DSM Biotechnology Center

Gordon, B., MIT / Broad Institute

Roubos, J. A., DSM Biotechnology Center

Voigt, C. A., Massachusetts Institute of Technology

Itaconic acid production by the natural producer *Aspergillus terreus* could be limited by difficult fermentation characteristics, such as high viscosity, shear stress, and low growth rate. An alternative host, such as*Saccharomyces cerevisiae*, might open new revenue for further process intensification. Various pathways for itaconic acid production in yeast, varying in the supply of AcCoA precursor, were designed using a genome scale metabolic model. By comparing the flux distribution between optimal growth and optimal production, additional targets were identified for overexpression and deletion. Four pathway variants were selected for â€˜buildâ€™ and â€˜testâ€™. The best tested variant was further improved in several iterative rounds using Design of Experiment methodologies. The full design-build-test-learn cycle delivered the highest itaconic acid titer in *S. cerevisiae* to date.

Genome Scale Metabolic Modeling and Analysis of Clostridium Difficile

Authors:

Norsigian, C. J., University of California

Palsson, B. Ø., University of California, San Diego

Monk, J. M., University of California, San Diego

Clostridium difficile is a pathogen of high clinical interest due to its persistence as a hospital-borne infection. Previous studies suggest that differences in metabolic capabilities are contributing factors to resistance and virulence. As a means of systematically understanding the linkage between genotype and phenotype of this organism we constructed a new genome-scale reconstruction of *C. difficile* strain 630 that builds and improves upon previous efforts. The model recapitulates experimental gene knockout predictions with 91% accuracy. We deploy constraint-based reconstruction and analysis along with flux-balance analysis to investigate metabolic capabilities by systematically predicting growth capabilities in over 180 nutrient environments. We utilize the reconstruction to build strain-specific models of 32 isolates including both symptomatic and asymptomatic strains and identified unique differentiating metabolic capabilities. We experimentally validate the models by generating Biolog phenotypic microarray data for all 32 of these strains and compared results to model predictions.

Genome Scale Metabolic Network Reconstruction of Thermotoga Sp RQ7

Authors:

Gautam, J., Bowling Green State University

One of the main challenges of developing alternative energy sources is the ability to compete economically with fossil fuels. One approach to make bioproduction more economical is to engineer organisms which can produce fuels as well as value added molecules. *Chromochloris zofingiensis,* a green alga, can accumulate up to 40% of their dry weight as triacylglycerols making it an excellent candidate for biofuels production. It also accumulates high amounts of astaxanthin, a high value ketocarotenoid that can be used as a pharmaceutical, nutraceutical, cosmetic, or as food or feed supplements[1-3]. Naturally derived astaxanthin has a market price of approximately $7,000 per kilogram. The co-production of this molecule product makes the economic feasibility of biodiesel production by *C. zofingiensis* much more attainable. In order to investigate the metabolic capacity of this organism for both fuel and astaxanthin production, we have reconstructed the metabolic network from the published genome. We will present our work to date and results of simulations to maximize the production of both products. (Research supported by grant DE-SC0018301 from the Department of Energy.)

References

1 Hussein, G., Sankawa, U., Goto, H., Matsumoto, K. & Watanabe, H. Astaxanthin, a Carotenoid with Potential in Human Health and Nutrition. *Journal of Natural Products* **69**, 443-449, doi:10.1021/np050354+ (2006).

2 Yuan, J. P., Peng, J., Yin, K. & Wang, J. H. Potential healthâ€□promoting effects of astaxanthin: A highâ€□value carotenoid mostly from microalgae.*Molecular Nutrition & Food Research***55**, 150-165, doi:doi:10.1002/mnfr.201000414 (2011).
3 Liu, J. *et al.* Chlorella zofingiensis as an Alternative Microalgal Producer of Astaxanthin: Biology and Industrial Potential. *Marine Drugs* **12**, 3487-3515, doi:10.3390/md12063487 (2014).

Genome-Scale Constraint-Based Metabolic Modeling and Analysis of Cryptoccocus Curvatus

Authors:

Pham, N., Wageningen University and Research
Suarez-Diez, M., Wageningen University and Research
Schaap, P. J., Wageningen Univeristy and Research
Martins Dos Santos, V. A. P., Wageningen University and Research
Reijnders, M., Wageningen University and Research
Nijsse, B., Wageningen University and Research

Cryptococcus curvatus, also known as*Cutaneotrichosporon curvatus*, an oleaginous fungus, is gaining attention due to its ability to use a wide range of carbon sources and its ability to accumulate more than 60% lipid in its biomass. In this study we constructed a genome-scale metabolic model for *C.curvatus* that can help to understand and achieve higher yield of lipid production.

A draft model was constructed using the scaffold-based method by Loira et al (2012). The draft model was extensively curated based on published data and literature with special emphasis on lipid synthesis pathways. The model was used to simulate growth on different carbon sources. RNAseq data were obtained at C:N of 16:1 and 8:5 with two replicates for each condition. These expression data were integrated into the model to study lipid synthesis in *C.curvatus*. The current version of the model has 831 genes, 1436 metabolites and 1679 reactions distribute in 11 compartments. The growth predictions are highly consistent with experimental data. However, the initial model was not able to predict differences in lipid synthesis at varying nitrogen concentrations. Integration of expression data will lead to improve predictions on lipid production at nitrogen abundant and nitrogen depleted conditions. These results illustrate that the model can be further used to study and guide metabolic engineering for optimization of*C.curvatus* as a cell factory for lipid production.

References

Loira, N., Dulermo, T., Nicaud, J.-M., & Sherman, D. (2012). A genome-scale metabolic model of the lipid-accumulating yeast Yarrowia lipolytica. *BMC Systems Biology*, *6*(1), 35. https://doi.org/10.1186/1752-0509-6-35

Genome-Scale Flux Elucidation in the Fast-Growing Cyanobacterium Synechococcus Elongatus Utex 2973

Authors:

Hendry, J. I., The Pennsylvania State University
Maranas, C. D., The Pennsylvania State University
Gopalakrishnan, S., The Pennsylvania State University
Pakrasi, H. B., Washington University
Tang, Y. J., Washington University

A short doubling time of 2h makes*Synechococcus* elongatus UTEX 2973 a promising platform for solar based production of biofuels and other bio-chemicals. Successful metabolic engineering strategies can be informed by the knowledge of intracellular flux distribution in this organism. In this meta-analysis, the intracellular flux distribution in a genome-scale metabolic model of *Synechococcus elongatus* UTEX 2973 was elucidated using isotopic non-stationary ^{13}C-metabolic flux analysis using the experimentally measured labeling dynamics of 13 central carbon metabolites reported in an earlier study. To this end, a genome scale carbon mapping model, *i*mSyu593, was constructed using the existing mapping model for *Synechocystis* sp PCC 6803, *i*mSyn617, as the starting point to trace the flow of carbons through 593 reactions spreading across the central carbon metabolism, amino acid metabolism and peripheral pathways. Flux elucidation revealed a near complete routing (>96%) of the assimilated carbons to biomass formation. This high biomass yield is afforded by the ability of this organism to reincorporate carbons oxidized in anabolic and photorespiratory pathways while increasing the flux through non-decarboxylating reactions. The ability to reincorporate oxidized carbons sustains a higher flux through the photorespiratory C2 cycle to meet the glycine and serine demand for the biomass synthesis. In contrast with other cyanobacteria, Malic enzyme is found to be dispensable with pyruvate being synthesized via Pyruvate Kinase. And instead of using the lower glycolysis and Pyruvate Dehydrogenase reaction, Acetyl-CoA was synthesized using the carbon efficient Phosphoketolase pathway. These findings suggest the existence of a highly efficient metabolism in *Synechococcus elongatus*UTEX 2973, which, in conjunction with fast growth rate, supports the development of this organism into an ideal photoautotrophic production platform.

Genome-Scale Metabolic Model of Chromochloris, an Emerging Model Organism for Sustainable Fuel Production

Authors:

Metcalf, A., Colorado School of Mines
Boyle, N. R., Colorado School of Mines

One of the main challenges of developing alternative energy sources is the ability to compete economically with fossil fuels. One approach to make bioproduction more economical is to engineer organisms which can produce fuels as well as value added molecules. *Chromochloris zofingiensis,* a green alga, can accumulate up to 40% of their dry weight as triacylglycerols making it an excellent candidate for biofuels production. It also accumulates high amounts of astaxanthin, a high value ketocarotenoid that can be used as a pharmaceutical, nutraceutical, cosmetic, or as food or feed supplements[1-3]. Naturally derived astaxanthin has a market price of approximately $7,000 per kilogram. The co-production of this molecule product makes the economic feasibility of biodiesel production by *C. zofingiensis* much more attainable. In order to investigate the metabolic capacity of this organism for both fuel and astaxanthin production, we have reconstructed the metabolic network from the published genome. We will present our work to date and results of simulations to maximize the production of both products. (Research supported by grant DE-SC0018301 from the Department of Energy.)

References

1 Hussein, G., Sankawa, U., Goto, H., Matsumoto, K. & Watanabe, H. Astaxanthin, a Carotenoid with Potential in Human Health and Nutrition. *Journal of Natural Products* **69**, 443-449, doi:10.1021/np050354+ (2006).
2 Yuan, J. P., Peng, J., Yin, K. & Wang, J. H. Potential healthâ€□promoting effects of astaxanthin: A highâ€□value carotenoid mostly from microalgae.*Molecular Nutrition & Food Research***55**, 150-165, doi:doi:10.1002/mnfr.201000414 (2011).
3 Liu, J. *et al.* Chlorella zofingiensis as an Alternative Microalgal Producer of Astaxanthin: Biology and Industrial Potential. *Marine Drugs* **12**, 3487-3515, doi:10.3390/md12063487 (2014).

Genome-Scale Metabolic Modeling of Actinomycetes for Secondary Metabolites Production

Authors:

Mohite, O., Technical University of Denmarl
Weber, T., Technical University of Denmark
Kim, H. U., Korea Advanced Institute of Science and Technology (KAIST)
Lee, S. Y., Korea Advanced Institute of Science and Technology (KAIST)

Actinomycetes are an important source for the discovery of medically useful secondary metabolites, such as anticancers and antibiotics. Major challenges of producing secondary metabolites include optimization of metabolic fluxes towards the biosynthesis of a target secondary metabolite and identification of an appropriate heterologous production host. Systems biology approaches including reconstruction of genome-scale metabolic models (GEMs) are increasingly applied to explore the dependencies between primary and secondary metabolism to predict gene manipulation targets for the overproduction of secondary metabolites. GMSM (Genome-scale metabolic Modeling with Secondary Metabolism) pipeline (unpublished) allows automatic reconstruction of GEMs with secondary metabolites biosynthetic pathways. . Here, we exemplify the strength of GMSM by reconstructing GEMs of important*a*ctinobacteria such as *Streptomyces venezuelae* and *Streptomyces collinus.*Growth prediction performance of GEMs under a variety of substrate conditions was compared with and fine-tuned using these two organismsâ€™

phenotype microarray data. The resulting GEMs can provide an important tool for applying systems metabolic engineering strategies for the optimal production of secondary metabolites in various actinomycetes as a heterologous production host. In this regard, GMSM can serve as a useful platform to explore the potential of various actinomycetes by generating a collection of GEMs to produce diverse secondary metabolites.

Genome-Scale Metabolic Reconstructions for Phylogeny?

Authors:

Schulz, C., NTNU - Norwegian University of Science and Technology
Almaas, E., NTNU - Norwegian University of Science and Technology

Constraint-based modeling is a widely used and powerful methodology to assess the metabolic phenotypes and capabilities of an organism. Genome-scale metabolic models, based on the genome and including all reactions of the organism, are mostly created for that purpose. However, that does not cover their full potential.
Since all reactions are included in the models, it is possible to compare these models and thereby the species directly by creating a consolidated reconstruction of several organisms. This consolidated reconstruction including every reaction occurring at least once allows for identifying cross-organism wide conserved reactions, linked to a gene of the source-organism.

AutoKEGGRec is an automated tool that creates first draft reconstructions of single organisms, community and a consolidated reconstruction for a list of organisms or strains. AutoKEGGRec is developed in Matlab and works seamlessly with the COBRA Toolbox v3, and it is based on only using the KEGG database as external input. The first draft reconstructions consist of all reactions for a KEGG organism ID and corresponding linked genes. This provides a comprehensive starting point for further refinement and curation using the host of COBRA toolbox functions or other preferred tools. Furthermore, additional functionalities are provided, like the organisms-reaction-genes matrix allowing for advanced cross-organism reaction engineering.

Here, a consolidated model for organisms across Domains was created using the innovative capabilities of AutoKEGGRec. The metabolic reconstructions were used to create a phylogenetic tree of life based on the basis of the reaction networks. Both the direct gene sequence as well as the underlying protein sequence were used.

This tree of â€œmetabolic networksâ€□ can be used to compare organisms focusing on the whole metabolome instead of using small parts of the DNA. Furthermore, it enlightens the metabolic engineering possibilities in between species throughout their maybe similar reactions for similar living environments in distinct points.

Grid-Based Computational Methods for the Design of Constraint-Based Parsimonious Chemical Reaction Networks to Simulate Metabolite Production: Gridprod

Authors:

Tamura, T., Kyoto University

Constraint-based metabolic flux analysis of knockout strategies is an efficient method to simulate the production of useful metabolites in microbes. Owing to the recent development of technologies for artificial DNA synthesis, it may become important in the near future to mathematically design minimum metabolic networks to simulate metabolite production. Accordingly, we have developed a computational method where parsimonious metabolic flux distribution is computed for designated constraints on growth and production rates which are represented by grids. When the growth rate of this obtained parsimonious metabolic network is maximized, higher production rates compared to those noted using existing methods are observed for many target metabolites. The set of reactions used in this parsimonious flux distribution consists of reactions included in the original genome scale model iAF1260. The computational experiments show that the grid size affects the obtained production rates. Under the conditions that the growth rate is maximized and the minimum cases of flux variability analysis are considered, the developed method produced more than 90% of metabolites, while the existing methods produced less than 50%. Mathematical explanations using examples are provided to demonstrate potential reasons for the ability of the proposed algorithm to identify design strategies that the existing methods could not identify. The source code is freely available, and is implemented in MATLAB and COBRA toolbox. This study has been accepted by BMC Bioinformatics.

Identification and Analysis of Bacterial Genomic Metabolic Signatures in the Department of Energy's Systems Biology Knowledgebase

Authors:

DeJongh, M., Hope College
Grimes, M., Hope College
Satkiewicz, L., Hope College
Tintle, N., Dordt College
Best, A., Hope College

With continued rapid growth in the number and quality of fully sequenced and accurately annotated bacterial genomes, we have an unprecedented opportunity to understand metabolic diversity. In previous work, we selected 101 diverse and representative completely sequenced bacteria and implemented a manual curation effort to identify 846 unique metabolic variants present in these bacteria. The presence or absence of these variants act as a metabolic signature for each of the bacteria, which can then be used to cluster them into a metabolic tree and analyze similarities and differences between and across bacterial groups. In our current work, we are developing an application to automate the assignment of metabolic variants to bacterial genomes in the Department of Energyâ€™s Systems Biology Knowledgebase (KBase). This will enable us to expand the set of bacteria to which this approach is applied and use the resulting tree to test broad questions about metabolic diversity and complexity across the bacterial tree of life.

This work is funded by the National Science Foundation awards MCB-1716285 to Hope College and MCB-1715211 to Dordt College.

References:

Nathaniel Bowerman, Nathan Tintle, Matthew DeJongh, Aaron A Best. Identification and Analysis of Bacterial Genomic Metabolic Signatures. Pacific Symposium on Biocomputing 2017, pp. 3-14. 2017.

AP Arkin, RW Cottingham, CS Henry et al. KBase: The United States Department of Energy Systems Biology Knowledgebase. Nature Biotechnology, Vol 36, pp. 566-569, 2018.

Identification of the Enzymes Responsible for Diverse Phenotypic States of Yeast Lipid Metabolism Using Comprehensive Mechanistic Models

Authors:

Tsouka, S., EPFL
Hatzimanikatis, V., Swiss Federal Institute of Technology (EPFL)

Lipids carry an important role in cell structure and function, as well as in the physiopathology of multiple diseases. Maintenance of the lipid profiles should be tightly regulated for preserving membrane permeability, cell integrity and several other functions.

Large-scale kinetic models of metabolic networks are essential in order to capture and predict such behaviors of cellular systems when subject to perturbations. To this end, we developed a detailed model of the lipid metabolism, based on genome-scale metabolic models of *S. cerevisiae*, in order to identify how the stoichiometric and kinetic coupling determines lipid homeostasis and its regulation.

We curated this model using thermodynamic data as well as lipidomic measurements and we used the Optimization and Risk Analysis of Complex Living Entities (ORACLE) framework to generate populations of parametrized kinetic models that are consistent with the given physiology, while satisfying the stoichiometric and thermodynamic constraints and accounting for the parametric uncertainty.

The model encompasses 1143 reactions and 799 metabolites across 6 cellular compartments (cytosol, mitochondria, peroxisomes, endoplasmic reticulum, Golgi and nucleus), and includes the following lipid-related subsystems: biosynthesis, elongation, and degradation of fatty acids, biosynthesis and esterification of sterols, biosynthesis of phospholipids, sphingolipids, and cardiolipin, triacylglyceride decomposition, dolichol biosynthesis and the mevalonate pathway. It also includes several key parts of yeast metabolism such as glycolysis, citric acid cycle, oxidative phosphorylation etc.

Using the distributions of the computed kinetic modelsâ€™ parameters, we constructed the dynamic mass balances of the species, in order to simulate the dynamic evolution of concentration profiles in response to small perturbations of enzyme activities, and to identify the enzymes that control the distributions of fluxes and metabolic concentrations at a representative steady state. We can further use this analysis to identify the changes in specific enzyme activities that are responsible for given mutant phenotypes.

Identifying Orphan Enzymes in Pseudomonas Putida Using Cobra-Based Methods

Authors:

Phan, B., University of Wisconsin-Madison
Vera Colon, C., University of Wisconsin-Madison
Bennett, N., University of Wisconsin-Madison
Pan, S., University of Wisconsin-Madison
Reed, J. L., University of Wisconsin-Madison

Genome-scale metabolic models have a wide variety of applications. While these models capture gene-protein-reaction relationships, they often are incomplete because they include misidentified reactions or metabolic gaps due to inaccurate genome annotations. Identifying genes responsible for missing reactions will provide a more complete understanding of metabolic networks and genotype-phenotype relationships. Previously, we developed a model-enabled gene search (MEGS) method to identify missing genes in a metabolic network associated with gap-filled reactions using functional selection experiments. MEGS involved creating a genomic library for a species with gap-filled reactions, and transforming the library into a recipient strain, such as *Escherichia coli*, whereby, the missing gap-filled reaction is coupled to growth [1]. Here, we extend MEGS to find genes associated with orphan enzymes in*Pseudomonas putida*. Databases exist that list orphan enzymes from many different organisms, whose metabolic activity has been shown experimentally but for which no associated gene sequences are known (including ~17% of enzymes with EC numbers) [2]. To verify some of the orphan enzymes in *P. putida*, we used COBRA methods to design *E. coli* recipient strains, which can only grow if they express genes associated with the appropriate orphan enzyme from *P. putida*. By finding sequences for orphan enzymes, we can build new GPR associations, thereby improving model predictions not only for *P. putida* but for other organisms with similar genes as well.

References:

1. Pan, S., Nikolakakis, K., Adamczyk, P. A., Pan, M., Ruby, E. G., & Reed, J. L. (2017). Model-enabled gene search (MEGS) allows fast and direct discovery of enzymatic and transport gene functions in the marine bacterium *Vibrio fischeri.Journal of Biological Chemistry,292*(24), 10250â€“10261.
2. Shearer, A. G., Altman, T., & Rhee, C. D. (2014). Finding sequences for over 270 orphan enzymes. *PLoS ONE, 9*(5).

Implementation and Application of a New Combined Stoichiometric and Thermodynamic Flux Balance Analysis

Authors:

Leupold, S., University of Groningen

Niebel, B., University of Groningen

Heinemann, M., University of Groningen

We recently uncovered that cellular metabolism is geared towards growth maximization but constrained by an upper rate of cellular Gibbs energy dissipation. Exploiting this principle in flux balance analysis generated excellent predictions of exchange and intracellular fluxes across a wide range of conditions and carbon sources, and even including some metabolite concentrations. This method requires a thermodynamically consistent formulation of cellular processes including Gibbs energies of reaction for every metabolic process, and a Gibbs energy balance, which states that the Gibbs energy dissipated by all cellular processes is equal to the Gibbs energy exchanged with the environment. We here present a detailed workflow, exemplified for a genome-scale metabolic model of *Escherichia coli,* how to implement and apply this method starting from any stoichiometric metabolic reconstruction. This workflow encompasses the formulation of such a model, the parameterization using experimental data together with regression analysis and the application in flux balance analysis. Furthermore, we present strategies how to solve the required nonlinear and nonconvex optimizations. Given the limited amount of required input data, and the precision and extent of the model predictions, we consider this method a valuable addition to current flux balance analysis approaches.

Implementing and Evaluating a Gaussian Mixture Framework for Identifying Gene Function from Tnseq Data

Authors:

Li, K., Columbia University
Chen, R., North Carolina State University
Lindsey, W., Dordt College
Best, A., Hope College
DeJongh, M., Hope College
Henry, C. S., Argonne National Laboratory
Tintle, N., Dordt College

The rapid acceleration of microbial genome sequencing increases opportunities to understand bacterial gene function. Unfortunately, only a small proportion of genes have been studied. Recently, TnSeq has been proposed as a cost-effective, highly reliable approach to predict gene functions as a response to changes in a cellâ€™s fitness before-after genomic changes. However, major questions remain about how to best determine whether an observed quantitative change in fitness represents a meaningful change. To address the limitation, we develop a Gaussian mixture model framework for classifying gene function from TnSeq experiments. In order to implement the mixture model, we present the Expectation-Maximization algorithm and a hierarchical Bayesian model sampled using Stanâ€™s Hamiltonian Monte-Carlo sampler. We compare these implementations against the frequentist method used in current

TnSeq literature. From simulations and real data produced by E.coli TnSeq experiments, we show that the Bayesian implementation of the Gaussian mixture framework provides the most consistent classification results.

Improving High Throughput Genome-Scale Metabolic Model Reconstruction and Validation with Tnseq Data Using Modelseed 2

Authors:

Faria, J. P., Argonne National Laboratory
Edirisinghe, J. N., Argonne National Laboratory
Liu, F., University of Minho
Seaver, S. M. D., Argonne National Laboratory
Jeffryes, J. G., Argonne National Laboratory
Zhang, Q., Argonne National Laboratory
Weisenhorn, P., Argonne National Laboratory
Sadkhin, B., Argonne National Laboratory
Gupta, N., Argonne National Laboratory
Gu, T., Argonne National Laboratory
Henry, C. S., Argonne National Laboratory

The Department of Energy Systems Biology Knowledgebase (KBase) is a platform designed to solve the grand challenges of Systems Biology. KBase has implemented bioinformatics tools that allow for multiple workflows including genome annotation, comparative genomics, and metabolic modeling. We selected a phylogenetically diverse set of approximately 1000 genomes and constructed draft genome-scale metabolic models (GSMMs) using the ModelSEED pipeline implemented in KBase. We used these 1000 genomes as a test set to improve the quality of models produced by the ModelSEED. First, we curated our mapping of RAST functional roles to biochemistry by reconciling with data mined from KEGG and published metabolic models; we corrected errors in our reaction reversibility assertions to improve overall model constraints; we applied a new method to predict auxotrophy across all 1000 genomes to predict improved gapfilling media; we refined our gapfilling procedure to prevent draft models from our pipeline from overproducing ATP; and we process all models through the Memote pipeline, accompanying complete reconstructions with Memote reports. We show how all of our pipeline improvements increase the number of gene associations, decrease the number of gapfilled reactions, improve the accuracy of growth and ATP production yield predictions, and decrease the number of blocked reactions across all models. The addition of Memote to our pipeline enables us to provide a measure for model quality that is consistent across reconstruction platforms. We show how auxotrophy, and pathway presence varies across our 1000 training-set genomes along the phylogenetic tree. We also plot model quality across the phylogenetic tree, identify taxa where model quality is lower. Finally, we select five specific genomes for which comprehensive TN-seq data is available, and we compare model predictions of all data with experimental results, showing significant improvement in accuracy between models generated by the original ModelSEED and models from ModelSEED 2.

Improving Succinic Acid Production in Actinobacillus Succinogenes through Genome-Scale Metabolic Models

Authors:

Alsiyabi, A., University of Nebraska - Lincoln
Long, D., University of Nebraska
Immethun, C., University of Nebraska
Saha, R., University of Nebraska-Lincoln

Identifying specific microbial drivers of variation in metabolic phenotypes is a major goal in the study of host-associated and environmental microbiomes. Correlation-based analysis of paired microbiome-metabolome datasets is a widespread approach to this objective. To date, however, the efficacy and limitations of this approach have not been evaluated. To address this challenge, we developed a mathematical definition of the contribution of each taxon to metabolite variation based on its uptake and secretion fluxes. We applied a multi-species dynamic genome-scale metabolic modeling pipeline to simulate simplified gut communities, generating idealized microbiome-metabolome datasets with precisely known microbial metabolic fluxes. Comparing the observed taxon-metabolite correlations in this simulated setting with the true taxonomic contributors, we found that correlation-based analysis poorly predicts key contributors, with low accuracy and a high false discovery rate. Importantly, however, the predictive value was strongly influenced by both metabolite and taxon properties, as well as exogenous environmental variation. These findings have practical implications for the analysis and interpretation of microbiome-metabolome studies. This study illustrates the utility of multi-scale metabolic modeling as a simulation tool to inform study design and methods development in microbiome research.

Kinetic Model of Clostridium Beijerinckii Based on Phenotypic States Superposition

Authors:

Rivas-Astroza, M., Pontificia Universidad Catolica de Valparaiso
Paredes, I., Pontificia Universidad Catolica de Valparaiso
Aroca, G., Pontificia Universidad Catolica de Valparaiso
Conejeros, R., Pontificia Universidad Catolica de Valparaiso

Growth kinetics follow lag, exponential, and stationary phases, with their transitions marked by drastic changes in biomass growth rates. This may be due to shifts in transcriptional activity, where high substrate concentrations promote increasing ribosome production to support the exponential phase. Under this hypothesis, there are smooth transitions from low to high growth rates on a phenotypically uniform population of cells. However, this is not the case for *Clostridium beijerinckii* which, when grown in batches, present various phenotypes coexisting. It starts as a vegetative phenotype, where fast substrate consumption is paired by fast biomass growth and production of acidic products. Acidification promotes a fraction of cells to murph into a clostridial phenotype. These are non-dividing cells whose metabolism is geared towards sporulation, a process where glucose and the previously released acidic species in the medium are consumed, producing in turn solvents. Released spores may later germinate into vegetative cells if conditions became favorable.

We propose a kinetic model where all three stages of *C beijerinckii* are present at any given time. The dynamics between phenotypic states are described by differential equations within dynamic FBA. To model the vegetative-clostrial and spore-vegetative transitions, we used the concentration of acidic species as arguments to a logistic equation whose value can be interpreted as the fraction of the vegetative cells transiting to clostridial state. Likewise, we used a second logistic equation, this time as a function of the concentration of solvents, to model the fraction of clostridial cells that change into spores.

Using public RNA-seq databases we determined divergently expressed genes in each phenotype. We pruned the metabolic network of C $beijerinckii$accordingly to create phenotype-specific metabolic networks. To train our model we measured changes in substrates, and acidic and solventogenic species along batch cultures of C $beijerinckii$ under various conditions.

Laboratory Evolution Reveals a Two-Dimensional Rate-Yield Tradeoff in Microbial Metabolism

Authors:

King, Z. A., University of California, San Diego
Cheng, C., University of California, San Diego

Growth rate (Âµ) and yield (Y) are fundamental features of microbial growth. However, we lack a mechanistic and quantitative understanding of the Âµâ€“Y relationship. Studies pairing computational predictions with experiments have shown the importance of maintenance energy and proteome allocation in explaining rate-yield tradeoffs and overflow metabolism. Recently, adaptive evolution experiments of*Escherichia coli* reveal a phenotypic diversity beyond what has been explained using a simple Âµâ€“Y relationship. Here, we identify a two-dimensional Âµâ€“Y tradeoff in adapted *E. coli* strains where the dimensions are (A) a tradeoff between Âµ and Y and (B) a tradeoff between substrate uptake rate (q_{glc}) and Y. We employ a multi-scale modeling approach, combining a previously reported small-scale proteome allocation model with a genome-scale model of metabolism and gene expression (ME-model), to develop a quantitative description of this two dimensional Âµâ€“Y relationship using data for *E. coli* K-12 MG1655. The analysis provides a mechanistic explanation of the two-dimensional Âµâ€“Y tradeoff. Furthermore, the analysis identifies modifications to the P/O ratio as a potential mechanism that enables the q_{glc}-Y tradeoff. Thus, the Âµâ€“Y tradeoffs that govern microbial adaptation to new environments are more complex than previously reported, and they can be understood in mechanistic detail using a multi-scale modeling approach.

Leveraging a Clostridium Difficile Genre and Metagenomics to Identify Candidate Probiotic Bacterial Strains

Authors:

Jenior, M., University of Virginia
Papin, J., University of Virginia

Clostridium difficile is a Gram-positive, sporulating anaerobe that has become the leading causing of hospital-acquired infection. Exposure to antibiotics sensitizes hosts to colonization of the gastrointestinal tract by *C. difficile* through disruption of the healthy resident bacterial community, known as the gut microbiota. Previous studies have strongly supported that colonization resistance to *C. difficile* is driven by competition for growth nutrients by members of the gut microbiota. As *C. difficile*is known to colonize numerous host species and is capable of catabolizing a large array of possible growth substrates, it is difficult to determine the bacterial groups in a resistant community that confer this property. In order to more closely study nutrient utilization by *C. difficile* during infection across distinctly susceptible environments within the complex milieu of the gut microbiome, we employed Flux-Balance Analysis in an *in vivo* contextualized GENRE of *C. difficile* strain 630. In consideration of large deficits in nutrient utilization pathways and in an effort to address several challenges with existing reconstructions of C difficile, we generated a *de novo* reconstruction of *C. difficile* 630 followed by extensive manual curation of its core metabolism and nutrient acquisition systems. *In silico* predictions in both rich and minimal medias, as well as transcriptomic and metabolomic data contextualization collected from an animal model of infection, closely mimic experimental characterization across multiple*in vitro* and *in vivo* studies. Furthermore, correlations between predicted *C. difficile*substrate utilization across distinct infection conditions with metagenome-enabled metatranscriptomic results from those microbiotas reveal discrete groups of bacteria which share distinct metabolic functionalities with the pathogen when compared to resistant communities. Ultimately, by identifying metabolic functionalities of resistant/sensitive communities we can design targeted probiotic therapies with the desired community phenotype.

Medusa: A Software Package for Construction and Analysis of Ensembles of Metabolic Networks

Authors:

Medlock, G. L., University of Virginia
Papin, J. A., University of Virginia

Genome-scale metabolic network reconstructions (GENREs) are powerful tools for predicting the metabolic behavior of organisms across broad environmental contexts and in the presence of drugs or genetic manipulation. Many applications of GENREs, such as contextualization of â€˜omics data and network expansion through integration of phenotypic data, lead to multiple feasible network structures which are all equally likely. Common practice is to choose a single network and perform downstream analyses without consideration of other equally-likely networks. Recently, maintaining these alternative network structures and performing analysis in an ensemble fashion has demonstrated improved predictive performance.
Here, we present Medusa, a software package for the construction and analysis of ensembles of GENREs. Medusa is an open-source python software package that performs constraint-based reconstruction and analysis (COBRA) methods at ensemble-scale. The underlying data structures implemented in Medusa extend cobrapy, a python package for analysis of GENREs using COBRA methods.

In this presentation, we highlight three key capabilities of Medusa. The first is generation and analysis of ensembles of GENREs from phenotypic data. The second is generation of ensembles composed of alternative GENRE states created using integration of transcriptomic data. The third key capability is identification of network components associated with variable predictions throughout an ensemble using supervised and unsupervised machine learning. Using this workflow of ensemble generation and uncertainty quantification, Medusa can be used to select experiments that maximally resolve uncertainty in GENRE structure.

Simulations performed using Medusa are probabilistic in that each ensemble generates a distribution of predictions. We hope that wide adoption of ensemble analysis methods will reduce investment in GENRE-based predictions that are spurious (e.g. only one ensemble member makes the prediction) and help guide experiments to quickly improve the quality and predictive power of GENREs.

Metabolic Analysis of Nitrogen Fixation in the Soybean Micro-Symbiont Sinorhizobium Fredii Using Constraint-Based Modeling

Authors:

Contador, C. A., University of California, Los Angeles
Wang, Q., Chinese University of Hong Kong
Chan, S. H. J., Colorado State University
Lam, H. M., Chinese University of Hong Kong

Rhizobia are soil bacteria known for being able to establish symbiosis with diverse host plants. In this context, *Sinorhizobium fredii*is a soil bacterium able to form nitrogen-fixing nodules with diverse legumes such as soybean. In particular, *S. fredii* strain CCBAU45436 represents one of the dominant sublineage of *S. fredii* that nodulates soybeans in alkaline-saline soils in the Huang-Huai-Hai Plain region in China. Here, we present a genome-scale metabolic model for the soybean micro-symbiont*Sinorhizobium fredii* strain CCBAU45436. The model was manually curated based on physiological, biochemical information and integration of high-throughput data. A symbiosis reaction was defined to represent the specific exchange of nutrients between the bacteroid and soybean. The symbiosis reaction together with the nitrogen fixation reaction defines the objective function to describe the stage of nitrogen fixation. The model developed can be used to systematically analyze symbiotic factors and predict essential genes for nitrogen fixation. In addition, the model can be expanded to represent different life stages of the bacteria to try to understand niche adaptation. No reconstructions are currently available for *S. fredii* strains despite of their potential benefits to sustainable agriculture and ecosystem.

This work is supported by the Hong Kong Research Grants Council GRF (14108014) and Area of Excellence Scheme (AoE/M-403/16) awarded to H.-M.L.

Metabolic Chaos during Microbial Stress Responses

Authors:

Jensen, P. A., University of Illinois at Urbana-Champaign

Bacteria mount a physiologic stress response to survive hostile or stringent conditions. The genes that comprise the stress response can be defined in two ways: a transcriptional network composed of genes with expression changes during stress, and a phenotypic network of genes whose deletion lowers the bacteria's fitness during stress. On first thought, one would expect these two networks to be the same, or at least have significant overlap. The genes important for surviving stress should be up-regulated, and genes not essential for tolerating the stress should be down-regulated. However, studies from multiple microbes and conditions have discovered that the transcriptional and phenotypic stress responses are nearly disjoint; genes with expression changes during stress can be deleted without changing fitness, and phenotypically important genes are rarely differentially expressed. Although several studies have described this striking phenomenon, it remains unclear why bacteria appear to have two separate stress response networks.

We discovered that fitness and expression changes can be linked using COBRA models. During nutrient depletion, genes with expression changes are often adjacent to phenotypically important genes. Even the magnitude of the fitness or expression changes are linked â€" the largest fitness changes are closest to the largest expression changes, and vice versa. The expression changes are much closer to the fitness changes than expected if both sets of genes were randomly distributed throughout the network. By contrast, the response to antibiotic stress is largely uncoordinated. Changes in fitness and expression appear randomly distributed across the network, and the magnitudes of the nearest fitness and expression changes are uncorrelated. Using a novel â€œenzyme couplingâ€□ framework, we show how modularity and redundancy in the metabolic network enable the split between transcriptional and phenotypic stress responses â€“ in both the coherent starvation response and the incoherent antibiotic response.

Metabolic Modeling of a Model Diatom

Authors:

Kumar, M., University of San Diego
Polson, S. W., University of Delaware
Allen, L. Z., J. Craig Venter Institute
Zengler, K., University of California

Diatoms are unicellular photosynthetic microalgae that are ubiquitous in fresh water and marine and environments. These microalgae are amenable for ~20% of the global carbon fixation. Diatoms are highly diverse and have substantially higher productivity and capabilities to accumulate lipids compared to other classes of microalgae. These properties make diatoms ideal candidates for advanced production of biofuels and chemicals. However, metabolism of diatoms is poorly characterized both in laboratory settings as well as in the environment, where diatoms interact. Previous studies revealed a very high abundance of marine viruses, some of which have been shown to infect diatoms along with other phytoplankton cells, and thus altering their metabolism. To gain in-depth knowledge about diatom metabolism and to address how viral infection modifies diatom physiology, we applied constraint-based modeling to the model diatom*Cylindrotheca* at different physiological states during its life-cycle. A genome-scale metabolic model of *Cylindrotheca* has been generated using annotated genome and three previously developed metabolic reconstructions of photosynthetic organisms (*Phaeodactylum tricornutum*,*Chlamydomonas reinhardtii*, and *Chlorella vulgaris*) as template database of biochemical reactions. The model is validated with growth/no growth physiological data at several growth conditions. Furthermore, the model is constrained using single-cell RNA-sequencing data with and without viral infection to elucidate the involved mechanisms in viral infection to*Cylindrotheca.* Metabolic model integrated with time-course gene expression data is employed to explore the physiological characteristics, such as growth and lipid biosynthesis at different stages, and to determine how viruses affect the metabolism of the diatom host at these states. The outcome of this work will underpin future studies to advance our understanding of diatom metabolism and to elucidate their role on the ecology and biogeochemistry of the ocean.

Metabolic Network Reconstruction for the Pan-Genome: A Scalable Method to Get High-Quality Metabolic Models across the Tree of Life

Authors:

Correia, K., University of Toronto
Mahadevan, R., University of Toronto

Background

Genome-scale metabolic models (GEM's) are a valuable tool used to study the metabolism of organisms with metabolic simulations, but they also provide a scaffold for 'omics data integration. The drop in cost for genome sequencing has lead to an exponential increase in the diversity of sequenced genomes, but the number of curated GEM's has not kept pace. This gap hinders our ability to study physiology across the tree of life. Furthermore, our review of published GEM's has found that metabolic models contain significant commission and omission errors in central metabolism. To address these quantity and quality GEM issues, we propose open and transparent efforts to curate the pan-genome, pan-reaction, and pan-metabolome of larger groups of organisms by research communities, rather than for a single species. We outline our approach for budding yeasts.

Method

We created a consensus metabolic network from 13 fungi/yeasts GEM's spanning seven species. We added additional reactions to the network to account for growth substrates and non-natural enzyme activities in the fungal literature that were not present in biochemical databases. Reactions were annotated with ortholog ID's from AYbRAH, our fungal ortholog database, to generate 33 fungi/yeasts GEM's from the subkingdom Dikarya.

Results

The fungal pan-GEM contains 2224 reactions, 2696 metabolites, 1543 orthologs, and 10 compartments. The gene coverage in the species models created from our pan-GEM is higher than published GEM's. Pathways not previously included in published GEM's, such as degradation of alkanes and aromatics, are captured using our approach.

Conclusion

Maintaining a metabolic network reconstruction annotated with a curated ortholog database at a higher pan-genome level improves the quality and quantity of GEM's. Capturing accurate ortholog-reaction-protein associations minimizes commission and omission errors, and the ability to scale GEM's to more branches of the tree of life.

Metabolic Network Reconstruction of Treponema Pallidum Spp. Pallidum

Authors:

Dräger, A., Center for Bioinformatics Tübingen (ZBIT)
Morini, S., Università degli Studi di Milano - Bicocca
Casini, I., University of Tübingen
Hamm, T. M., Center for Bioinformatics Tübingen (ZBIT)
Nieselt, K., University of Tübingen

Background: Since the discovery, made in 1905 by Schaudinn and Hoffmann, of Treponema pallidum ssp. pallidum as the etiologic agent of syphilis, medicine has made significant progress against this disease. Yet, despite the availability of diagnostic tests and a therapy based on antibiotics, the world has not stopped being burdened by syphilis, that has been re-emerging globally over the last few decades (WHO Report, 2008), for which no vaccine is still available and which, moreover, in its early stages enhances the transmission of HIV. Continuous in vitro culture of this organism has still not been achieved, imposing a substantial roadblock to its experimental inspection, and even the sequencing of its genome (Fraser et al., 1998) did not yield an obvious solution to the cultivation problem. While much has already been tried on the laboratory bench, this pathogen has still not (to our knowledge) been tackled using a systems biology approach.

Results: Here, we present a first manually curated draft reconstruction of the metabolic network in Treponema pallidum ssp. pallidum towards a genome-scale metabolic model (GEM). At this time, the model iSM161 comprises 161 genes of 1,039 predicted open reading frames that are responsible for 239 reactions of 277 metabolites. The model is still under development and steadily updated. For the reconstruction, COBRApy has been used, where subsystem information is added and parsed as SBML groups extension using libSBML.

Discussion: Using this reconstruction, together with the application of COBRA methods, we anticipate to gather new insights into the pathogenâ€™s physiology and pathology, and in how this spirochete, which has earned the designation of â€œstealth pathogen,â€ succeeds in making a living and eluding humanâ€™s immune defenses as well as cultivation attempts. It is planned to make the model available to the community in SBML format.

Metabolic Supply and Demand Constrain Optimal Regulatory Strategies: Structural Evidence from the E coli Small Molecule Regulatory Network

Authors:

Euler, C., University of Toronto
Mahadevan, R., University of Toronto

Balancing supply and demand is a life or death-or-task for bacteria. In resource-limited envrionments, they must be able to precisely coordinate metabolic flux distributions to balance the demands required for optimal growth without wasting costly metabolites. To this end, they have evolved complex, layered regulatory systems which respond to changing external contexts. While the structure and function of hierarchical regulatory mechanisms (transcription and translation) have been characterized extensively, regulation at the metabolic level remains less well understood. To address this, we reconstructed a genome-scale small-molecule regulatory network (SMRN) for E coli to understand the role small-molecule regulation plays in balancing supply and demand. Structural analysis of our SMRN revealed a dense network of interactions with layers of organization. We found that at the local level, SMR interactions are organized into well-understood feedforward loop motifs that provide specific input-output relationships between fluxes through relatively distanct metabolic reactions. Functionalization of these motifs demonstrates that the SMRN has the capacity to influence supply and demand balancing in two ways: by coordinating signals within the core metabolism and effecting tertiary fluxes based on these signals; and by directly balancing fluxes of metabolites which are essential for biomass production. We explore the constraints under which this regulatory system is likely to be active to show that it can work synergistically with transcriptional regulation and that protein degradation rate is the dominant factor which determines the optimal balance of control burden across these layers of regulation.

Mgpipe: A Toolbox for Investigating Metagenomics Data through Personalized Microbiota Metabolic Models

Authors:

Baldini, F., University of Luxembourg
Thiele, I., University of Luxembourg
Heinken, A., University of Luxembourg
Fleming, R. M. T., Leiden University
Heirendt, L., University of Luxembourg

mgPipe: A toolbox for investigating metagenomics data through personalized microbiota metabolic models

Background/Purpose: Predictive, personalized medicine requires tools that can account for individual variability, such as the genome, life style, and the microbiome. Recently, the gut microbiome has received great attention for its potential to modulate human metabolism and its role in health and disease. The aim of this project is the development of an easy-to-use tool (mgPipe (1)) that enables researchers to investigate *in silico* the functional implications associated with changes in gut microbial composition in healthy individuals and patients.

Methods: mgPipe combines published algorithms, which map metagenomics data onto a reference database to determine microbial identity and abundance (2-3). Subsequently, this microbial information is combined with genome-scale metabolic reconstructions of more than 800 gut microbes (4) to generate personalized microbiome metabolic models. Using established constraint-based modeling methods from the COBRA toolbox (5) and machine learning approaches, mgPipe enables the in silico interrogation of the different functional, metabolic properties of these microbiome models.

Results and conclusion: mgPipe is an efficient tool to integrate metagenomic data with constraint based modeling enabling creation, simulation, and analysis of personalized microbiome metabolic models.

References

1. Baldini F, Heinken A, et al. The Microbiome Modeling Toolbox: from microbial interactions to personalized microbial communities. bioRxiv. 2018:318485.
2. Bolger AM, et al. Bioinformatics. 2014:btu170.
3. Li H, et al. Bioinformatics. 2009;25(14):1754-60.
4. Magnúsdóttir S, et al. Generation of genome-scale metabolic reconstructions for 773 members of the human gut microbiota. Nature Biotechnology. 2016.
5. Heirendt L, Arreckx S, et al. Creation and analysis of biochemical constraint-based models: the COBRA Toolbox v3. 0. arXiv preprint arXiv:171004038. 2017.

Mighty Morphing Metabolic Models: Leveraging Manual Curations for Automatic Metabolic Reconstruction of Clades

Authors:

Simeonidis, E., Institute for Systems Biology
King, B., Institute for Systems Biology
Richards, M. A., Institute for Systems Biology
Price, N. D., Institute for Systems Biology

Genome-scale metabolic models have been employed with great success for phenotypic studies of organisms over the last two decades. The most difficult step in the reconstruction of these models is manual curation; a time-intensive and laborious process of literature review that is nevertheless essential for a high-quality network. Although various methods have been developed to accelerate automated reconstruction, much effort must still be expended to supplement an automatically-generated draft model with manually curated information from the published literature to achieve the quality needed for successful simulation of the metabolic processes of the organism. Here, we are utilizing our tool for likelihood-based gene annotation (King *et al.*, 2018) to create a method that â€œmorphsâ€□ a manually curated metabolic model to a draft model of a closely related organism. Our method combines genes from the original, manually curated model with genes from an annotation database to create a final structure that contains gene-associated reactions from both sources. The benefits of such an approach are twofold: 1) the effort and accumulated knowledge that has gone into the construction of the original model is leveraged to create a metabolic model for a closely related organism; 2) starting from an already completed and functioning model allows the user to run simulations at every step as necessary, offering the ability to predict how modifications will affect the performance of the model. We demonstrate our method by applying it on our model of *Methanococcus maripaludis* (Richards *et al.*, 2016), to create morphed models of three related methanogenic archaea. Model morphing offers a viable alternative to other automated reconstruction methods, particularly for organisms that are evolutionarily dissimilar to those that form the foundation of annotation databases, and provides a faster method to reconstruct a clade of metabolic models for related organisms from a manually curated representative.

Miren: An Optimization Tool for Data-Driven Discovery of Global Regulatory Phenomena Used to Elucidate the Heat Stress Response Mechanism in Rice Seed

Authors:

Islam, M. M., University of Nebraska-Lincoln
Sandhu, J., University of Nebraska-Lincoln
Walia, H., University of Nebraska-Lincoln
Saha, R., University of Nebraska-Lincoln

Plants use a suite of strategies to respond to abiotic stresses, including changing the abundance of stress-responsive genes/proteins that ultimately lead to the large-scale changes in gene expression levels, protein abundances, and metabolites. Complex gene-protein-reaction associations as well as regulatory mechanisms constitute a challenge to elucidate stress response mechanisms in plants. This research aims to understand the heat stress response mechanisms in developing rice seed using temporal transcriptomic analyses in control and stress conditions and *in silico* metabolic analyses.

Rice plants were exposed to heat stress at 12 and 36 hours after fertilization with a 16h-light/8h-dark cycle, and young developing seeds were collected from control and stressed plants. Total RNA isolated from developing seeds was used for differential gene expression analysis, which yielded ~7000 significantly stress-responsive genes. Clustering analysis was used to develop a minimal gene interaction network and identify global regulators. The highly connected â€œhubâ€ genes included previously-identified MADS-box genes as well as a large number of novel regulatory genes. MiReN, an MILP optimization-based tool was developed to decipher the minimal regulatory network using the time-series transcriptomic data. MiReN identified important regulatory relationships for stress-responsive rice transcription factors (*e.g.*, OsMYB, OsbZIP, OsMADS etc.) and predicted the minimal global regulatory network for rice seed in control and stress conditions. A comparative analysis of the network topology reveals the shift in regulatory mechanisms in presence of stressors and allows for integration of transcriptomic data with a genome-scale metabolic model of rice seed. Work on other rice tissues and modeling the interactions between them using multi-level and multi-objective modeling frameworks to develop a robust plant-scale rice model is underway. Our predictive mathematical model will identify biologically important and non-intuitive solutions to questions related to stress response mechanisms and accelerate the development of tolerant plant varieties in an efficient and accurate fashion.

MLProScape: Machine Learning (ML) Based Method for Engineering Enzymes Faster By Modeling Protein Fitness Landscape (ProScape)

Authors:

Gupta, S. T. P., Great Lakes Bioenergy Research Center
Glasgow, E., University of Wisconsin Madison, Madison, WI
Fox, B. G., University of Wisconsin Madison
Ramanathan, P., University of Wisconsin Madison
Reed, J., University of Wisconsin Madison

Protein engineering aims to improve the functional properties of a protein, such as thermo-stability, binding affinity, and/or catalytic activity and has become a vital step in developing industrial enzymes. This work describes a machine learning (ML) based method called *â€˜MLProScapeâ€™* that builds an accurate model of the protein fitness landscape (ProScape) and then, uses the model to design synthetic protein designs with superior functional properties. Unlike approaches using directed evolution, which requires experimentally screening millions of protein variants, *MLProScape* requires fewer protein variants (on the order of tens to hundreds) to be tested experimentally, owing to the power of statistical inference.

As a proof-of-concept, *MLProScape* was applied to enhance the catalytic activity of glycoside hydrolases â€“ a key enzyme used to degrade lignocellulosic biomass for biofuel production. Experimentally measured specific activities for a diverse set of glycoside hydrolases were used to train the ML models. The resulting elastic net regression models have a high predictive power (with correlation coefficient and R^2 values as high as 0.896 and 0.714, respectively, between the predicted and experimentally measured specific activities using a 5-fold cross validation). Moreover, by using position specific features, amino acid positions distal to the active site that might play a key role in modulating the activity level can be identified. *MLProScape* is also capable of modeling complex design criteria, such as engineering the catalytic activity of an enzyme towards multiple substrates simultaneously, as well as, to account for other desirable traits such as high stability and better *in vivo*expression.

Development of methods like*MLProScape* will complement and add value to the current growing repertoire of *in-silico* pathway engineering tools as it will enable metabolic engineers to alleviate bottleneck steps en route target chemical of interest.

Key words: *machine learning, enzyme engineering, sequence-to-function, experimental design*

Model-Based Prediction of Yersinia Enterocolitica Infection Outcome

Authors:

Dräger, A., Center for Bioinformatics Tübingen (ZBIT)
Geißert, J., University Hospital Tübingen
Eichner, M., Epimos GmbH
Bohn, E., University Hospital Tübingen
Mostolizadeh, R., University of Tübingen
Autenrieth, I. B., University Hospital Tübingen
Beier, S., Center for Bioinformatics Tübingen (ZBIT)
Schütz, M., University Hospital Tübingen

The Lactobacillus genus has, primarily, dominated the probiotic industry with a great deal of success in conferring health benefit as shown in a variety of rigorous clinical trials. In this study, we generate over 200 genome-scale metabolic reconstructions from a broad array of Lactobacillus species, including many of the currently sequenced species that are typically found in marketed probiotics, resident strains from human isolation sites, and fermented foods. We have conducted a strain level comparison of the reactomes for each genome, demonstrating that the metabolic reactome contributes to clustering of the genomes based on human health related isolation sites, including the gut, oral cavity, and vagina. Strains resident to the vaginal microbiota cluster separately from strains resident to the gastrointestinal tract, while there is a large overlap in metabolic potential between gut and oral residents. A random forest analysis was used to identify reactions specific to resident strains to the gastrointestinal tract when compared to other lactobacillus strains. Carbon and amino acid source utilization was predicted using the generated metabolic reconstructions, our data indicate that there is a great deal of variation among Lactobacillus strains with some correlation to isolation site. We present a phylogenetic analysis based on metabolic capabilities, demonstrating the similarities among Lactobacillus strains indicating which groups are most likely to compete in a microbial community. The metabolic reconstructions were generated using ModelSEEDs universal format, the PATRIC database, Mackinac, CobraPy, and Probannopy. This study provides a public resource of systematically generated genome scale metabolic reconstructions for a wide range of Lactobacillus strains for further curation and analysis of probiotics relevant to human oral, gut, and vaginal health. Additionally, this work lays the ground work for personalize design, selection, and delivery of Lactobacillus based probiotics.

Model-Guided Engineering of Cyanobacteria for Improved Biofuel Production

Authors:

Purdy, H. M., University of Wisconsin-Madison
Reed, J. L., University of Wisconsin—Madison

Cyanobacteria are promising hosts for fuel and chemical production as their main metabolic inputs are minimal and renewable (i.e. light, carbon dioxide, water, and a nitrogen source). However, cyanobacteria engineered to produce compounds of interest frequently suffer from low yields and/or are genetically unstable. As such, metabolically coupling chemical production to growth may be important in engineering cyanobacteria because it can generate robust production characteristics. Additionally, a growth-coupled production approach may help to circumvent gaps in our knowledge about cyanobacterial metabolic regulation, which likely factors into the poor production characteristics that are often reported relative to heterotrophic strains. In this work, a genome-scale metabolic model is being used to guide engineering of the cyanobacterium*Synechococcus* sp. PCC 7002 for growth-coupled overproduction of short- to mid-chain alcohols (e.g. *n*-butanol) that are potential biofuels. PCC 7002 is of particular interest as a cyanobacterial production strain due to its relatively rapid growth-rate and tolerance to saline and high-light conditions. Metabolic engineering strategies for improving the production of various alcohols in PCC 7002 were investigated via the metabolic model *i*Syp708, which was previously developed by our lab. Using this approach, a potential strategy for high-yield growth-coupled production of *n*-butanol and 2-methyl-1-butanol in PCC 7002 was successfully identified. This strategy hinges on coupling production of these alcohols to nitrate assimilation by rewiring PCC 7002â€™s native NADH-cycling pathways. Strains are currently being constructed to test this approach. If a strongly growth-coupled production strain is successfully engineered, it will be adaptively evolved to further improve its alcohol production characteristics and genetic changes in these evolved strain(s) will be identified and analyzed. This study demonstrates that genome-scale metabolic models are useful in guiding cyanobacterial metabolic engineering and will help identify genetic alterations for optimizing the production of biofuel compounds in cyanobacteria.

Modeling of Potentially Virulence-Associated Metabolic Pathways in Pseudomonas Aeruginosa PA14 Including Experimental Verification

Authors:

Dräger, A., Center for Bioinformatics Tübingen (ZBIT)
Renz, A., University of Tübingen
Bohn, E., University Hospital Tübingen
Schütz, M., University Hospital Tübingen

According to the report of the â€˜Antimicrobial resistance surveillance in Europe (2015),â€™ *P. aeruginosa* is an opportunistic, human pathogen that causes many infections in hospitalized patients with immune defects or impairments. Since it is difficult to control *P. aeruginosa* in hospitals, it can cause hospital-acquired pneumonia [1]. Some strains of *P. aeruginosa* seem to be associated with higher mortality than others. With the help of a published genome-scale model of PA14 [2] and sequencing data of a highly pathogenic patient strain that was recently isolated in TÃ¼bingen, metabolic differences between the laboratory and patient strain shall be identified and subsequently verified in a laboratory experiment.

First, differences in the sequences of the two strains were identified by performing SNP analysis. These differences were then used to find metabolic alterations affecting the virulence. By using FBA and *in-*

*silico*gene knock-out, three genes were identified that could be responsible for the difference in pathogenicity between the laboratory and the highly pathogenic patient strain. These genes affect metabolic reactions that are associated with virulence in *P. aeruginosa*. Gene knock-outs of these three genes are momentarily being performed in a laboratory experiment to verify their metabolic relevance in virulence.

However, only a fraction of the genes of *P. aeruginosa* is included in the published model. Many additional genes are associated with virulence, differ between the sequenced laboratory and patient strains, or both. Therefore, the model is being extended to increase its predictive value.

1. European Centre for Disease Prevention and Control. Antimicrobial resistance surveillance in Europe 2015. Annual Report of the European Antimicrobial Resistance Surveillance Network (EARS-Net). Stockholm: ECDC; 2017.https://ecdc.europa.eu/en/publications-data/antimicrobial-resistance-surveillance-europe-2015#no-link
2. Bartell JA, Blazier AS, Yen P, Thøgersen JC, Jelsbak L, Goldberg JB, and Papin JA. Reconstruction of the metabolic network of *Pseudomonsas aeruginosa*to interrogate virulence factor synthesis, Nature Communications (2017).doi:10.1038/ncomms14631

Modeling Suboptimal Community Growth and Enhanced Species Diversity of the Gut Microbiota

Authors:

Henson, M. A., University of Massachusetts Amherst
Phalak, P., University of Massachusetts Amherst

Underlying cellular responses is a transcriptional regulatory network (TRN) that modulates gene expression. Current descriptions of the TRN cannot quantitatively deconvolute the transcriptome into the relative contributions of individual transcriptional regulators. Here, we applied unsupervised learning to a compendium of high-quality *Escherichia coli* RNA-seq datasets (median $R^2 = 0.99$ between biological replicates) to identify 71 statistically independent regulatory signals. Summation of the 71 signals explained over 80% of expression variation across 115 unique experimental conditions. Of these, 50 signals were directly linked to characterized transcriptional regulators. Condition-specific signal strengths were validated by exposure to new environmental conditions, confirming 76% of predicted signal activations. The resulting decomposition of the transcriptome provided: 1) a quantitative, mechanistic explanation of responses to environmental and genetic perturbations, 2) a guide to gene function discovery, 3) characterization of a novel pyruvate-responsive transcription factor, and 4) a basis for comparing transcriptional regulation across closely related strains. Thus, we find that signal summation forms an underlying principle that can describe the composition of the prokaryotic transcriptome.

Modelling of Cell Metabolism to Reduce the Lactate Generation in CHO Cell Cultures

Authors:

Martínez-Monge, I., Technical University of Denmark
Volkova, S., Technical University of Denmark
Marín de Mas, I., Technical University of Denmark
Hefzi, H., University of California
Comas, P., Universitat Autònoma de Barcelona
Lewis, N., University of California
Lecina, M., IQS-Universitat Ramon Llull
Cairó, J. J., Universitat Autònoma de Barcelona
Nielsen, L. K., Technical University of Denmark

Chinese Hamster Ovary cells (CHO) display Warburg metabolism characterized by high glucose consumption and high lactate production under aerobic conditions. Lactate is a by-product widely reported to inhibit cell growth in culture, imposing an important burden to industrial processes. In order to reduce lactate secretion and thus its inhibitory effects, two different approaches have been applied: i) Bioprocess Engineering, controlling extracellular conditions in the bioreactor to trigger concomitant glucose/lactate consumption, and ii) Synthetic Biology, to generate non-lactate producer mutant cells knocking out lactate dehydrogenase and the regulators responsible for inhibiting pyruvate conversion to acetyl-CoA.

In the current study, the metabolic profile of CHO cells was investigated in batch bioreactor cultures performed under three conditions: normal

WT, controlled WT, and zero-lactate CHO (CHO ZeLa). Exometabolomic data was integrated into a reduced genome-scale metabolic model using Flux Balance Analysis (FBA) during the mid-exponential phase, as well as Dynamic FBA (DFBA) to capture the dynamic changes occurring over time in CHO cell metabolism. Model reduction was performed through a novel semi-automated reduction protocol to generate fully functional metabolic models.

FBA showed that in wild-type cell line, lactate is produced to fulfill the NADH regeneration requirements in the cytoplasm and only a small amount of pyruvate is introduced into TCA through Acetyl-CoA. When concomitant glucose and lactate consumption was triggered, as well as in CHO ZeLa, glucose uptake was significantly reduced and a balance between glycolysis and TCA cycle fluxes was reached, yielding a more efficient substrate consumption. Moreover, DFBA illuminated the metabolic mechanisms by which wild-type CHO switches from a Warburg (glucose consumption/lactate production) phenotype to a glucose/lactate co-consumption phenotype.

Our approach enabled us to explore the mechanisms underlying dynamic metabolic response of wild-type CHO and CHO ZeLa with potential implications in the industry of bioproducts.

Models Constrained with Transcriptomics or Proteomics Data Generate Discordant Predictions

Authors:

Carey, M. A., University of Virginia
Untariou, A., University of Virginia
Papin, J. A., University of Virginia

Modern high-throughput technologies have enabled detailed characterizations of an organismâ€™s phenotype; this large-scale molecular data allows for a systems-level analysis of biological conditions. Moreover, genome-scale metabolic reconstructions (GENREs) can be leveraged to interpret these data by integrating â€˜omics into a GENRE, resulting in a condition-specific models and the generation condition-specific predictions. Transcriptomic data have been most frequently used for data integration due to the datasetsâ€™ capacity for genome-wide coverage. However, regulation occurs at the translational level as well as transcriptional level, so proteomics-derived models may be used to generate more accurate condition-specific predictions than transcriptomic-derived models.

To investigate the functional differences between expression integration using either transcriptomic and proteomic data, matched datasets from five biological conditions (GSE65209 and PXD001659) were integrated into a GENRE (from Bosi,*et al.,* doi: 10.1073/pnas.1523199113) using GIMME. As transcriptomic datasets have higher coverage than many proteomics datasets, the effect of data coverage was investigated by random sampling subsets of the data. Predictions (specifically, reaction essentiality) from models generated with matched transcriptomic and proteomic data (*i.e.* collected from the same biological condition) and mismatched datasets (*i.e.* from different conditions) were compared; surprisingly, these matched predictions were no more similar than predictions generated from mismatched models. Transcriptomic-derived models predicted more essential reactions than proteomics models, even when controlling for scope (*i.e.* number of enzymes constrained), indicating transcriptomic-derived models are more constrained than proteomics-derived models. These results emphasize the importance of interpreting predictions as condition-, datatype-, and approach-specific results. Future work will explore the effects of alternative integration approaches and datasets.

Multiomics Integration for Prediction of Complex Metabolic Phenotypes

Authors:

Zelezniak, A., The Francis Crick Institute

A key challenge in solving genotype to phenotype relationship is to predict a cellâ€™s metabolome. Here, we quantified the proteomes in 97 non-essential kinase knock-out strains from *Saccharomyces cerevisiae*and associated these to their metabolomes. We observed that kinase knockouts affect proteomes broadly but in a distinct manner; and that these are quantitatively dominated by expression changes in metabolic enzymes. Analysing these data in the context of kinetic modeling demonstrated that enzyme abundance affects metabolite concentrations through the redistribution of flux control, resulting in a many-to-many relationship between enzyme abundances and the metabolite concentrations. Machine learning enabled mapping these relationships, the prediction of experimentally measured metabolite concentrations, as well as to identify candidate genes important for the regulation of metabolism. Overall, our study suggests that hierarchical metabolism regulation acts predominantly through adjustment of broad expression patterns rather than over individual rate-limiting enzymes, and may account for more than half of metabolism regulation.

Multiple Reconstructions of P. Putida for Metabolic Engineering Applications

Authors:

Mueller, J., University of Nebraska - Lincoln

Pseudomonas putida is an industrially relevant species of bacteria. It has Generally recognized as safe (GRAS) status, is amenable to genetic manipulations, and has a robust metabolism and tolerance to several solvents used in industry. These attributes give it the potential to be a versatile platform for the bio-industry. Since 2003 multiple different strains of *P. putida* with varying capabilities have had their genomes fully sequenced. Previously detailed models of the strain KT2440 have been created, but little has been done about other strains. By creating a multi-strain reconstruction of more than 20 strains of this species, new pathways and usages can be elucidated. A pan- and core-genome can be identified which will aid in identifying central pathways as well as peripheral pathways that can be manipulated to add new capabilities and improve output.

New Methods to Constrain Genome Scale Models with Stable Isotope Labeling Data

Authors:

Backman, T., Joint BioEnergy Institute
Ando, D. E., Joint BioEnergy Institute (JBEI)
Keasling, J. D., Technical University of Denmark
Garcia Martin, H., Joint BioEnergy Institute (JBEI)

Here we present new methods to directly constrain genome scale models with 13C stable isotope labeling data. By reducing the assumptions required for flux modeling, these methods can infer fluxes in engineered organisms with unusual phenotypes. First, we present a method to systematically calculate flux bounds for any specified â€œcoreâ€□ of a genome-scale model so as to satisfy the bow tie approximation that carbon tends to flow outward from core metabolism. Second, we automatically identify an updated set of core reactions that can satisfy this approximation more efficiently via Simulated Annealing. Lastly, we present a workflow to quantify the uncertainty of fluxes constrained with stable isotope labeling data at the genome scale using Markov Chain Monte Carlo (MCMC).

Optfill: A Novel Optimization-Based Tool to Automate the Gapfilling of Genome-Scale Metabolic Models

Authors:

Schroeder, W., University of Nebraska - Lincoln
Saha, R., University of Nebraska-Lincoln

Computational modeling of metabolism is now an indispensable tool to drive the processes of understanding, discovering, and redesigning of biological systems. By defining the metabolic space, genome-scale metabolic (GSM) models can assess allowable cellular phenotypes and explore metabolic potential under specific environmental and/or genetic conditions. GSM model curation processes typically involve gleaning information on gene annotations and reactions from major public databases; however, incomplete gene annotations and system knowledge leaves gaps in any GSM reconstruction. Gapfill (and numerous similar tools) automates addressing of these gaps, applying Mixed Integer Linear Programming (MILP) and utilizes functionalities from related organisms or changing the direction of existing reactions to fill gaps on a per-gap basis. This is done without consideration of thermodynamically infeasible cycle (TIC) avoidance. Hence, Gapfill makes redundant changes and increases the number of TICs in GSM models, which ultimately require further manual scrutiny. To address the limitations of current automated gapfilling procedures, introduced here is an improved method, namely OptFill, which performs automated gapfilling. OptFill applies two MILP problems in series, which addresses the fixes needed on a per-GSM model basis. The first optimization problem (single-level MILP) seeks to identify all TICs which could be created by adding new functionalities to the GSM model. The second optimization problem (multi-level MILP) seeks to maximize the number of gap metabolites fixed subject to minimizing the number of new functionalities added. Included in the second problem is an integer cut based on the results of the first problem, which prevents any solution containing TICs from being viable solutions. OptFill has thus far been successfully applied to metabolic models of a generic test system, *Saccharomyces cerevisiae* (bakerâ€™s yeast), *Exophiala dermatitidis*. Thus, OptFill provides a distinct advantage over the traditional Gapfill approach in the extent of automation and needed manual curation.

Predicted Responses of a Large-Scale Pseudomonas Putida KT2440 Kinetic Metabolic Model to Several Single-Gene Knockouts Are Consistent with Experimental Observations

Authors:

Miskovic, L., EPFL
Tokic, M., Swiss Federal Institute of Technology (EPFL)
Hatzimanikatis, V., Swiss Federal Institute of Technology (EPFL)

P. putida emerged as one of the most promising production hosts for a wide range of chemicals, due to its fast growth with a low nutrient and cellular energy demand, considerable metabolic versatility, ability to grow in wide range of chemicals, suitability for genetic manipulations and its robustness and high flexibility to adapt and counteract different stresses. One of the main advantages of *P. putida* compared to commonly used industrial hosts such as *E. coli* is its superior tolerance to toxic compounds such as benzene, toluene, ethylbenzene, xylene and other hydrocarbons. In this work, we developed a large-scale kinetic model of *P.putida* to predict the metabolic phenotypes and design metabolic engineering interventions for the production of biochemicals. We first performed a gap-filling and thermodynamic curation of the genome-scale iJN1411 model of *P. putida* KT2440. The redGEM and lumpGEM algorithms for the systematic reduction of stoichiometric genome-scale models are then applied to the curated iJN1411 to derive a consistently reduced large-scale stoichiometric model of *P. putida*. Using this model as a scaffold, we next employed the ORACLE framework to generate a population of large-scale kinetic models around the experimentally observed steady state. To illustrate the predictive capabilities of these models, we performed two studies. First, for a wild-type strain of *P. putida* KT2440 growing on glucose under aerobic conditions, we computed metabolic responses to several single-gene knockouts, and the developed kinetic models successfully captured the experimentally observed phenotypes. In the second study, we proposed metabolic engineering interventions for improved robustness of this organism to stress conditions. Overall, the results from these studies suggest that the developed models of *P. putida* metabolism can successfully be used for metabolic engineering design.

Predicting Shifts in Cardiomyocyte Metabolism during Heart Failure

Authors:

Dougherty, B., University of Virginia
Rawls, K., University of Virginia
Kolling, G., University of Virginia
Kalyan, V., U.S. Army Medical Research and Materiel Command
Pannala, V., U.S. Army Medical Research and Materiel Command
Wallqvist, A., U.S. Army Medical Research and Materiel Command
Papin, J. A., University of Virginia

Contextualizing the human metabolic network to model the metabolic states of specific tissue- and cell-types has generated novel hypotheses and tremendous insight into the role of metabolism in disease. Here, we use a recently published human GENRE, iHsa, to model cardiomyocyte metabolism through the integration of proteomics data available in the Human Protein Atlas (HPA). We validate the cardiomyocyte model through completion of pre-defined cardiomyocyte relevant tasks as well as evaluation of general tasks published with the original human metabolic network reconstruction. Both sets of tasks have provided a starting point for manual curation of the cardiomyocyte model. Next, we use the cardiomyocyte model to identify reactions, pathways, and genes differentially affected in heart failure through the utilization of different carbon sources for ATP production, an established phenotype in heart failure. The cardiomyocyte model demonstrates enrichment for carbon-source specific pathways, demonstrating the ability of the model to connect genotype to phenotype using the phenotypic changes observed in carbon substrate utilization that occur in heart failure.

Probabilistic Metabolic Network Reconstruction: A Modular Approach to Integrate Omics Data and Represent Uncertainty

Authors:

Bernstein, D., Boston University
Ofaim, S., Boston University
Zoccarato, L., Leibniz-Institute of Freshwater Ecology and Inland Fisheries
Sher, D., University of Haifa
Segrè, D., Boston University

Metabolic network reconstruction is central to many aspects of metabolic analysis and engineering, and is essential for ongoing efforts towards mechanistic microbial ecosystem modeling. In general, the reconstruction process involves translating genomic information into a metabolic network that can provide insight into cell-wide metabolic fluxes and other phenotypes. A number of well-curated reconstructions exist for different model organisms, but a major challenge in the field is to reconstruct metabolic networks for diverse environmentally and medically relevant organisms. Various data sources including genomic, metabolomic, and phenotypic data can be used to guide metabolic network reconstruction. Additionally, there are several methods available for automatically generating draft reconstructions, including by using probabilistic genome annotations. However, few methods have directly attempted a comprehensive, automated integration of multiple data types, while at the same time explicitly dealing with the inherent uncertainty in the reconstruction process. We have developed a Markov chain Monte Carlo method to sample metabolic networks from a probability space shaped by the available data. Our method formalizes the data integration process and clarifies the representation of uncertainty by generating a natural ensemble of metabolic network reconstructions. As a proof of concept, we implemented this method on simulated genomic and metabolomic data from the *E. coli* iJO1366 metabolic network. Moving forward, we are working on including additional data types, such as phenotypic assays, with the goal of applying this method to a large group of environmentally-relevant marine microbes.

Reconciling Barseq Datasets for Z. Mobilis Using a Genome-Scale Metabolic Model

Authors:

Courtney, D., Chemical and Biological Engineering, University of Wisconsin, Madison, WI
Pan, S., University of Wisconsin-Madison
Reed, J., University of Wisconsin Madison
Ong, W. K., University of Wisconsin-Madison

Zymomonas mobilis is an industrially relevant, gram negative, ethanologen known for very high glycolytic fluxes, high ethanol production, and low biomass yields. Development of engineered strains of *Z. mobilis* has been relatively slow, and future efforts stand to benefit from being informed and directed by updated genome-scale metabolic models. Here we present iZM4_472, an updated model of *Z. mobilis*ZM4, and use it to analyze and reconcile BarSeq datasets from pooled mutant fitness experiments [Deutschbauer et al. *Journal of bacteriology* (2014)]. iZM4_472 contains 757 metabolic and transport reactions (of which 611 have GPR associations), 472 genes, and 632 unique metabolites, making it the largest and best annotated model of *Z. mobilis* to date. Data from reported pooled mutant experiments was used to assess the accuracy of gene essentiality predictions and identify genes associated with gap-filled reactions in the model. Several discrepancies between gene essentiality predictions and experimental results were caused by gene duplication, polar effects, or mis-mapped barcodes in the mutant library; thus, highlighting potential challenges with applying these high-throughput datasets to improve metabolic models. Correlations between transposon mutant fitness across 492 experiments were used to identify gene candidates for gap-filled reactions involved in histidine biosynthesis. Additional genes for reactions involved in tyrosine, biotin, ubiquinone, and pyridoxal 5â€™-phosphate synthesis were identified and confirmed through a combination of *E. coli* mutant complementation experiments, and our previously developed model-enabled gene search (MEGS) approach [Pan et al. *Journal of Biological Chemistry* (2017)]. Finally, flux coupling analysis of iZM4_472 was used to analyze the fitness scores of all 492 experiments in the reported BarSeq datasets to identify metabolic modules where mutant phenotypes were poorly correlated. These modeling efforts help pinpoint where additional metabolic knowledge gaps of *Z. mobilis* metabolism reside.

Representation of ME-Models in Sbml

Authors:

Dräger, A., Center for Bioinformatics Tübingen (ZBIT)
Voigt, M. A., University of Tübingen
Lloyd, C. J., University of California, San Diego
Yang, L., University of California, San Diego
King, Z. A., University of California, San Diego
Kohlbacher, O., University of Tübingen
Nieselt, K., University of Tübingen

Metabolism and Expression models (ME-models) [1] are a constraint-based modeling approach, which explicitly accounts for the cost of macromolecular biosynthesis. This allows ME-models to investigate genotype-phenotype relationships with a quantitative incorporation of '-omics' data. Common standards are a prerequisite for interoperability of systems biology tools. Novel approaches often require additional or changed data structures that cannot be represented in existing standards. This has been the case for ME-models. ME-models are powerful tools, but a lack of standards for encoding and creation, and the increased complexity of these models prevented widespread use.

SBMLme is an extension of current model encoding standards that enables SBML representations of ME-models that were reconstructed using COBRAme [2]. A prototype of the extension has been created in Java together with a standalone, bi-directional converter, between this extension and COBRAme's model storage format. The converter showed that SBMLme could fully and correctly encode a COBRAme model.

The use of SBMLme enables ME-models to be shared more efficiently, and a wider variety of tools to access ME-models. This should promote the propagation of ME-models. SBMLme may be used as a proof-of-concept for an official SBML package for ME-models. The standardization process requires ongoing discussion, the identification of a consensus SBML draft standard, and the implementation of the existing validation rules.

SBMLme is freely available athttps://github.com/draeger-lab/SBMLme under the terms of the MIT license.

1. Ines Thiele, Ronan M. T. Fleming, Richard Que, Aarash Bordbar, Dinh Diep, and Bernhard O. Palsson. Multiscale Modeling of Metabolism and Macromolecular Synthesis in E. coli and Its Application to the Evolution of Codon Usage. PLOS ONE, 7(9):1-18, 09 2012

2. Colton J Lloyd, Ali Ebrahim, Laurence Yang, Zachary Andrew King, Edward Catoiu, Edward J O'Brien, Joanne K Liu, and Bernhard O Palsson. COBRAme: A Computational Framework for Models of Metabolism and Gene Expression. bioRxiv, 2017.

Sammi: An Interactive, Semi-Automated Tool for the Illustration and Visualization of Metabolic Networks

Authors:

Schultz, A., University of Texas MD Anderson Cancer Center
Akbani, R., University of Texas MD Anderson Cancer Center

The visualization of large metabolic networks is a challenging process. Bottom-up, manual construction of such maps is extremely laborious, while automated methods can yield useful but often muddled networks. Therefore, the development of semi-automated network visualization tools, which automatically generate large-scale metabolic maps while at the same time allowing user-defined manual curation, are necessary. Here we present SAMMI, an interactive tool for the illustration and visualization of metabolic networks. SAMMI draws metabolic networks as dynamic bipartite graphs. Force-directed node orientation allows for the constant repositioning of network nodes, assisting in the map design and visualization process. SAMMI also provides a wide array of customization features that allow users to efficiently build metabolic maps. These functionalities include positioning, fixing, grouping, splitting, and coloring of network nodes, temporarily removing secondary metabolites, adding and removing metabolites and reactions, adding annotation text and shapes, and parsing the model into subgraphs based on model annotation (e.g. reaction subsystem or metabolite compartment) or user-defined input (e.g. lists of reactions). The implementation of graph theory algorithms also allow for the fast selection and re-arrangement of relevant network nodes. SAMMI illustrations can be exported in SAMMI specific or ESCHER compatible formats, as well as PDF images suitable for publications. We believe SAMMI is a powerful tool for the illustration of metabolic network maps, as well as the visualization of multiple data types.

Sbmltools: A Sbml Integration Tool for Kbase

Authors:

Liu, F., Argonne National Laboratory
Faria, J. P., Argonne National Laboratory
Zhang, Q., Argonne National Laboratory
Rocha, M., University of Minho
Rocha, I., University of Minho
Henry, C. S., Argonne National Laboratory

Since the release of the first version of System Biology Markup Language (SMBL) it become the de facto exchange format to store and share Genome Scale Metabolic Models (GSM). However, due to limitations in earlier versions of the SBML users adopted their own strategies to represent important attributes regarding to GSMs: gene-protein-reaction rules (GPR), metabolite/reaction annotation and model constraints. While most of these features were addressed in the latest version of SBML it left a heterogenous community of SBML models.

The SBMLTools provides methods to import and integrate existing SBML models into the Department of Energy Systems Biology Knowledgebase (KBase). The purpose of the SBMLTools is to reshape SBML models to fit the KBase demands: separation of the exchange reactions for the media formulation, integration of the metabolic network with the ModelSEED biochemistry, integration of GPR genes to match with a compatible genome and normalize the compartments with the vocabulary in KBase.

The tool compiles an internal in-house built integrated biochemistry database that compiles existing compound and reaction databases (ModelSEED, KEGG, MetaCyc, BiGG, etc) for annotation purposes. It also takes advantage of the KBase genome catalog that hosts more than 70.000 RefSeq bacterial genomes, indexing several attributes regarding to their genes (e.g., gene aliases).

While the main purpose of the SBMLTools is to improve the compatibility of the existing tools in KBase with external built models, it can also be used by researchers to normalize SBML models regarding to identifiers (compounds and reactions), exchange media, compartments and genes.

Selective Metabolic Vulnerabilities in Multiple Myeloma

Authors:

Valcárcel, L. V., CIMA, University of Navarra
Planes, F. J., TECNUN, University of Navarra
Apaolaza, I., TECNUN, University of Navarra
Agirre, X., CIMA, University of Navarra
Prosper, F., CIMA, University of Navarra
Ordoñez, R., CIMA, University of Navarra
Meydan, C., Cornell University
Melnick, A., Cornell University

Multiple myeloma (MM) is a hematological cancer characterized by a heterogeneous clinical presentation and an abnormal accumulation of clonal plasma cells in the bone marrow. MM is the second most common cancer worldwide and accounts more than 30,000 new diagnoses cases in 2016 just in United States. Immunomodulatory drugs and proteasome inhibitors are the main existing treatments against MM, which significantly improves the life expectancy of patients; however, MM remains an incurable disease.

In order to identify novel therapeutic strategies in MM, we systematically search for selective metabolic vulnerabilities using the concept of genetic Minimal Cut Sets (gMCSs), recently introduced in Apaolaza et al. 2017. To that end, we first enumerated 20,000 gMCSs from Recon3D (Brunk et al. 2018), the most recent reconstruction of the human metabolism. Secondly, RNA-seq data from the CoMMpass (Relating Clinical Outcomes in Multiple Myeloma to Personal Assessment of Genetic Profile) project, the largest study performed to date by the Multiple

Myeloma Research Foundation (MMRF), was mapped onto the gMCSs to discover metabolic targets in MM. Third, in order to identify selective target for MM cells, we also analyzed an unpublished RNA-seq study that includes healthy samples for different cell types arising from the B-cell differentiation: bone marrow plasma cells, tonsil plasma cells, memory B cells, centrocytes, centroblasts, naïve B cells. Fourth, the same analysis was repeated in different MM cell lines in order to identify suitable candidates for in-vitro experimental validation. Main results and future directions are presented and discussed.

[1] Apaolaza, Iñigo, et al. "An in-silico approach to predict and exploit synthetic lethality in cancer metabolism." Nature communications 8.1 (2017): 459.

[2] Brunk, Elizabeth, et al. "Recon3D enables a three-dimensional view of gene variation in human metabolism." Nature biotechnology 36.3 (2018): 272.

System-Level Examination of Metabolism in Rhizosphere Grown Burkholderiales

Authors:

Navid, A., Lawrence Livermore National Laboratory

Pett-Ridge, J., Lawrence Livermore National Laboratory

Star, E., University of California

Banfield, J. F., University of California, Berkeley

Firestone, M., University of California

Nuccio, E., Lawrence Livermore National Laboratory

It is well known that rhizosphere (in the vicinity of plant roots) microbes have a significant effect on plant health and growth rate. However, our knowledge of the system-level biochemical interactions between these microbes and their surroundings is incomplete. To gain a better understanding of these interactions, we have sequenced the soil metagenomes and constructed metagenome assembled genomes (MAGs) of active organisms growing in rhizosphere and bulk soil. We have developed a genome-scale model of metabolism for one of these MAGs (a member of the Burkhoderiales family, SM_S39) and used the model to explore time-resolved gene-expression data for samples collected within and outside of the influence of roots.

Our results show that SM_S39 greatly upregulates its import and metabolism of inorganic and complex nitrogen sources such as amino acid dimers, chitosamine when growing in the rhizosphere. In addition, this organismâ€™s rhizosphere CAZyme transcripts were upregulated for enzymes potentially involved in the decomposition of cellulose, hemicellulose, cell membranes, lipids, polysaccharides, proteins, starches, and biopolymers.

We also identified pathways that were significantly upregulated when comparing the bulk soil to the rhizosphere soil environment. We found that pathways related to generation and consumption of fatty metabolites and those involved in biosynthesis and metabolism of amino acids (especially aromatic, branched-chain, as well as some common nitrogen carrying amino acids aspartate and arginine) were strongly upregulated. These results indicate that although it is intuitive to think that exchange of carbon between the root and the microbes is the primary driver of the metabolic interactions between the plant and the soil microbes like SM_S39, microbial nitrogen metabolism might be affected even more significantly by the presence of roots.

Systematic Reduction of Genome-Scale Models for the Study of Metabolic Phenotypes of Human Cells

Authors:

Masid, M., EPFL
Ataman, M., Swiss Federal Institute of Technology (EPFL)
Hatzimanikatis, V., Swiss Federal Institute of Technology (EPFL)

In the last years, the analysis of cellular metabolism has sparked new interest in systems biology and metabolic modelling. In particular, modelling the phenotypes of healthy and diseased cells will help to understand the main metabolic characteristics of disease development and progression. It will also be key to design more effective therapies.

The reconstruction of genome-scale models (GEMs) enables the computation of phenotypic traits based on the genetic composition of a target organism. To overcome the well-known challenges when working with large networks, we generate systematically reduced models around specific subsystems. Within this framework, we curate the GEMs to include the thermodynamic properties of the network metabolites and reactions. Furthermore, we consider the composition and utilization of the extracellular-medium metabolites, and the synthesis of the biomass precursor metabolites. The reduced models can be used for a broad range of applications extending from *omics*data integration to kinetic models. We demonstrate here that the study of reduced human GEMs can provide insight and a systematic framework to compare different versions of

GEMs (Recon 2 versus Recon 3) as well as different cellular physiologies, such as, diseased versus healthy phenotypes.

Systems Approaches to Identify Metabolite Signatures in Placental Development

Authors:

Baloni, P., Institute for Systems Biology

Price, N. D., Institute for Systems Biology

Paquette, A., Institute for Systems Biology

Brockway, H., Cincinnati Children's Hospital Medical Center

Sadovsky, Y., Magee-Womens Research Institute

Muglia, L., Cincinnati Children's Hospital Medical Center

The human placenta is essential for successful pregnancy as it performs critical functions such as transfer of nutrients and oxygen to the fetus, removing harmful waste products, and providing mechanical, hormonal and immune support. Alterations to placental function can have profound impacts on the health of the developing fetus and the outcome of the pregnancy. In this study, we analyzed transcriptomic data of 200 term placentae and identified expression levels for metabolic genes. We used the most comprehensive human metabolic reconstruction, Recon 3D, and integrated transcriptomics data with the model to make it placenta-specific. The draft reconstruction of placenta consists of about 5525 reactions, more than 5500 metabolites and captures about 80% reactions in subsystems such as Vitamin K metabolism, cholesterol metabolism, citric acid cycle, chondroitin synthesis, nucleotide metabolism and glycerophospholipid metabolism compared with Recon 3D model. This is the first *in silico* metabolic network of the human placenta and will serve as a platform to integrate longitudinal transcriptomic and metabolomic data throughout pregnancy. We have also developed a pipeline to integrate a placental transcriptional regulatory network with this metabolic network, thus making it possible to predict transcription factors that regulate metabolic changes in placenta. Placenta metabolic reconstruction provides insights into normal placental development and can also be used to predict metabolic signatures in pregnancy complications. We aim to develop a predictive methodology that combines information from various omics measurements with metabolic reconstruction that can be used to predict a metabolic trajectory across pregnancy.

Tailoring a Metabolic Model for CHO Cell-Line By Integrating Data-Inferred Metabolic Activities

Authors:

Schinn, S. M., University of California, San Diego
Wei, W., Pfizer
Xu, Y., Pfizer
Morrison, C., Pfizer
Zhang, L., Pfizer
Lewis, N., University of California

CHO cells are widely used in commercial production of biotherapeutics. Despite this, CHO bioprocesses have traditionally been developed empirically, due to a limited understanding of their biology. Now, an increased knowledge of CHO cell biology can be exploited to improve production and control. A promising platform for such rational engineering is genome-scale metabolic network models, which describe metabolism mathematically. These models can predict cell-line specific metabolic features when integrated with omics data, based on key assumptions on gene expression regulation and gene-to-reaction relationships. Here, we deployed a novel approach that uses transcriptomic data to define a cells' metabolic functions, and then used these to build a metabolic model for a CHO-K1-derived industrial producer cell-line. We have used the resulting model to predict key reaction fluxes for multiple bioprocess settings, and it can be applied for rational bioprocess optimization.

Template of Metabolic Reprogramming in Cancer and Healthy Cells for Inferring Oncogenes

Authors:

Wang, F. S., National Chung Cheng University
Wu, W. H., National Chung Cheng University
Li, F. Y., National Chung Cheng University
Shu, Y. C., National Chung Cheng University
Lai, J. M., Fu-Jen Catholic University
Chang, P. M. H., National Yang Ming University
Huang5, C. Y. F., National Yang Ming University

Cancer cells exhibit unusual metabolic activity, characterized by high rates of glucose consumption and lactate production even under aerobic conditions, as well as increased glutamine catabolism and amino acid metabolism. Based on the behaviors of metabolic reprogramming, oncogenesis can be inferred from a genome-scale metabolic model of cancer cells. In this study, we incorporated the CORDA method [1] with data of the Human Protein Atlas database [2] and Recon 2.2 network [3] to reconstruct tissue-specific metabolic models for normal and cancer cells, respectively. These models were applied to assess the flux alternation template using flux-sum balance analysis. We developed a bilevel optimization formulation to infer oncogenes for the tissue-specific models. The objective function in the outer optimization problem is the similarity ratio of flux alternations of the dysregulated cell to the template for cancer or health models, or Warburg hypothesis from the literature. The inner optimization problem consisted of the flux balance model and uniform flux distribution model [4]. A nested hybrid differential evolution algorithm was introduced to solve the bilevel optimization problem. The head and neck tissue-specific model was used for a case study to infer oncogenes. Results show that TPK1 achieved the highest similarity ratio of 0.84 and three genes (GNPD1/GNPD2, THTPA, and PTEN) exhibited similarity ratio of 0.839. Literature shows that TPK1 is an oncogene for colon cancer [5] and PTEN is a tumor suppressor gene of squamous cell carcinomas of the head and neck [6].

1. Schulta et al., PLOS Computational Biology, 12(3): e1004808, 2016
2. https://www.proteinatlas.org/
3. Swainston1 et al., Metabolomics, 12:109, 2016
4. Wu et al., PLOS Computational Biology, 13(7): e1005618, 2017
5. Tiwana1 et al., Oncotarget, 6(8) : 5978â€“89, 2015
6. M. Poetsch et al., Cancer Genetics and Cytogenetics, 132(1):20-4, 2002

The genetic basis for adaptation of model-designed syntrophic co-cultures

Authors:

Lloyd, C. J., University of California, San Diego
Ebrahim, A., University of California, San Diego
King, Z. A., University of California, San Diego
Yang, L., University of California, San Diego
O'Brien, E. J., University of California, San Diego
Liu, J., University of California, San Diego
Catoiu, E., University of California, San Diego
Palsson, B. O., University of California, San Diego

Understanding the fundamental characteristics of microbial communities has far reaching implications for human health and applied biotechnology. However, the genetic basis of viable community formation of synthetic communities has not been studied in detail. By pairing auxotrophic mutants in co-culture, it has been demonstrated that viable nascent E. coli communities can be established where the mutant strains are metabolically coupled. A novel algorithm, OptAux, was constructed to design 61 unique multi-knockout E. coli auxotrophic strains that require significant metabolite uptake to grow. These predicted knockouts included a diverse set of novel non-specific auxotrophs that result from inhibition of major biosynthetic subsystems. Three OptAux predicted non-specific auxotrophic strainsâ€"with diverse metabolic deficienciesâ€"were co-cultured with an L-histidine auxotroph and optimized via adaptive laboratory evolution (ALE). Time-course sequencing revealed the genetic changes employed by each strain to achieve higher community growth rates and provided insights into mechanisms for adapting to the syntrophic niche. A community model of metabolism and gene expression was utilized to predict the relative community composition and fundamental characteristics of the evolved communities. This work presents new insight into the genetic strategies underlying viable nascent community formation and a novel computational method to elucidate metabolic changes that empower the creation of cooperative communities.

The Jsbml Project: A Fully Featured Java API for Working with Systems Biological Models

Authors:

Dräger, A., Center for Bioinformatics Tübingen (ZBIT)
Rodriguez, N., The Babraham Institute,
Hamm, T. M., University of Tübingen
Schulte, R., University of Tübingen
Watanabe, L., University of Utah
Vazirabad, I. Y., Marquette University
Kofia, V., University of Toronto
Myers, C. J., University of Utah
Hucka, M., The California Institute of Technology

With continued rapid growth in the number and quality of fully sequenced and accurately annotated bacterial genomes, we have an unprecedented opportunity to understand metabolic diversity. In previous work, we selected 101 diverse and representative completely sequenced bacteria and implemented a manual curation effort to identify 846 unique metabolic variants present in these bacteria. The presence or absence of these variants act as a metabolic signature for each of the bacteria, which can then be used to cluster them into a metabolic tree and analyze similarities and differences between and across bacterial groups. In our current work, we are developing an application to automate the assignment of metabolic variants to bacterial genomes in the Department of Energyâ€™s Systems Biology Knowledgebase (KBase). This will enable us to expand the set of bacteria to which this approach is applied and use the resulting tree to test broad questions about metabolic diversity and complexity across the bacterial tree of life.

This work is funded by the National Science Foundation awards MCB-1716285 to Hope College and MCB-1715211 to Dordt College.

References:

Nathaniel Bowerman, Nathan Tintle, Matthew DeJongh, Aaron A Best. Identification and Analysis of Bacterial Genomic Metabolic Signatures. Pacific Symposium on Biocomputing 2017, pp. 3-14. 2017.

AP Arkin, RW Cottingham, CS Henry et al. KBase: The United States Department of Energy Systems Biology Knowledgebase. Nature Biotechnology, Vol 36, pp. 566-569, 2018.

The Strain Optimization API: A Flexible Strain Design Formalism for an Automated High-Throughput Industrial Pipeline

Authors:

Ng, C. Y., The Pennsylvania State University

As a pioneer in making molecules in a sustainable way by strain engineering and industrial fermentation, Amyris has invested heavily in technology to develop strains and bring valuable molecules to the market at unprecedented speed. In the past three years, we have developed the Automated Scientist, a suite of mathematical models and machine learning algorithms that can iteratively perform the Design-Build-Test-Learn cycle for automated strain design and improvement. Thus far, the Automated Scientist has successfully created > 50,000 strains producing over 190 molecules. The design module of the Automated Scientist incorporates genome-scale metabolic models, an in-house knowledge store of design elements annotated with public data, and proprietary design algorithms to create pathway and strain designs. The pathway design to a target molecule can be visualized in the open-source tool Escher (first extended for industrial application at Amyris in 2014 by the author of the code) for route curation and additional flux balance simulation by biologists. To facilitate the exchange of knowledge between human scientists and the Automated Scientist, we have recently developed the Strain Optimization API. The API allows human scientists to specify the intended strain designs at a high level of abstraction, to validate their designs *in silico*, and to visualize the genotype-phenotype relationships. The Automated Scientist can parse the high-level input and automatically design the genetic architecture of the strains. In addition, the API can be accessed by Amyrisâ€™ high-throughput omics pipeline to identify strain-specific gene/protein/metabolite targets for testing up to 5,000 samples/day, and thus seamlessly accelerate the strain re-design process.

The Systems Biology Graphical Notation: A Standardised Representation of Biological Maps

Authors:

Dräger, A., Center for Bioinformatics Tübingen (ZBIT)
Touré, V., Norwegian University of Science and Technology
Mazein, A., Université de Lyon
Rougny, A., National Institute of Advanced Industrial Science and Technology
Dogrusoz, U., Bilkent University
Luna, A., Harvard Medical School

With continued rapid growth in the number and quality of fully sequenced and accurately annotated bacterial genomes, we have an unprecedented opportunity to understand metabolic diversity. In previous work, we selected 101 diverse and representative completely sequenced bacteria and implemented a manual curation effort to identify 846 unique metabolic variants present in these bacteria. The presence or absence of these variants act as a metabolic signature for each of the bacteria, which can then be used to cluster them into a metabolic tree and analyze similarities and differences between and across bacterial groups. In our current work, we are developing an application to automate the assignment of metabolic variants to bacterial genomes in the Department of Energyâ€™s Systems Biology Knowledgebase (KBase). This will enable us to expand the set of bacteria to which this approach is applied and use the resulting tree to test broad questions about metabolic diversity and complexity across the bacterial tree of life.

This work is funded by the National Science Foundation awards MCB-1716285 to Hope College and MCB-1715211 to Dordt College.

References:

Nathaniel Bowerman, Nathan Tintle, Matthew DeJongh, Aaron A Best. Identification and Analysis of Bacterial Genomic Metabolic Signatures. Pacific Symposium on Biocomputing 2017, pp. 3-14. 2017.

AP Arkin, RW Cottingham, CS Henry et al. KBase: The United States Department of Energy Systems Biology Knowledgebase. Nature Biotechnology, Vol 36, pp. 566-569, 2018.

Thermodynamic Metabolic Flux Analysis with Consistent Handling of Errors in the Estimates from the Component Contribution Method

Authors:

Mahamkali, V., Australian Institute for Bioengineering and Nanotechnology
Valgepea, K., The University of Queensland
McCubbin, T., University of Queensland
Marcellin, E., The University of Queensland
Nielsen, L. K., Australian Institute for Bioenegineering and Nanotechnology
Thermodynamic metabolic flux balance analysis (TMFA) is used extensively in genome scale metabolic modeling to obtain thermodynamically consistent flux and metabolic concentration profiles. TMFA involves the use of thermodynamic constraints to avoid thermodynamically infeasible solutions and provides information about free energy change of reactions as well as the range of metabolite activity. The current best method for estimating standard Gibbs energy of reaction (Î"$_r$G^0) uses the component contribution method combining reactant and group contribution methods. The estimates of Î"$_r$G^0 are associated with an error, which is calculated from errors on the individual reactant and group contributions.

Recent implementations of TMFA allow for the error in the estimates by allowing each individual Î”$_r$G^0 to vary independently within its estimated confidence. However, this is not the correct treatment as the individual Î”$_r$G^0 are not independent; they share reactant and group contributions. Allowing them to vary independently will inevitably result thermodynamic inconsistencies, where closed loops have non-zero Î”G. This problem was resolved by returning to the original TMFA formulation, where the error was allowed in the individual groups and then identically propagated to all Î”$_r$G^0. Using the component method, this approach is here expanded by allowing for errors in both reaction and group contributions (each is allowed with their respective confidence ellipsoid) and then propagating the adjusted estimates correctly to all Î”$_r$G^0. The method is made available as an OpenCobra function.

Thermodynamics As an Optimization Goal for Metabolism: Prediction of Metabolite Levels, Rate Constants and Post-Translational Regulation

Authors:

Cannon, W. R., Pacific Northwest National Laboratory
Zucker, J. D., Pacific Northwest National Laboratory
Baxter, D. J., Pacific Northwest National Laboratory
Kumar, N., Pacific Northwest National Laboratory
Dunlap, J. C., Geisel School of Medicine at Dartmouth
Hurley, J. M., Rensselaer Polytechnic Institute
Baker, S. E., Pacific Northwest National Laboratory

Nature selects those organisms that can reproduce the fastest while maintaining fitness and extracting the least amount of energy from their environment. This problem can be formulated as a maximum entropy production rate problem that includes experimental and physical constraints. We report the application this approach for modeling biological systems in lieu of having*in vivo*rate constants. The method is applied in four steps: (1) a new constrained optimization approach based on Marcelinâ€™s 1910 mass action equation is used to obtain the maximum entropy distribution, (2) the predicted metabolite concentrations are compared to those generally expected from experiment using a loss function from which post-translational regulation of enzymes is inferred, (3) the system is re-optimized with the inferred regulation from which rate constants are determined from the metabolite concentrations and reaction fluxes, and finally (4) a full ODE-based, mass action simulation with rate parameters and allosteric regulation is obtained. From the last step, the power characteristics and resistance of each reaction can be determined. The method is applied to the central metabolism of *Neurospora crassa*and the flow of material through the three competing pathways of upper glycolysis, the non-oxidative pentose phosphate pathway, and the oxidative pentose phosphate pathway are evaluated as a function of the NADP/NADPH ratio. It is predicted that regulation of phosphofructokinase (PFK) and flow through the pentose phosphate pathway are essential for preventing an extreme level of fructose 1, 6-bisphophate accumulation. Such an extreme level of fructose 1,6-bisphophate would otherwise result in a glassy cytoplasm with limited diffusion, dramatically decreasing the entropy and energy production rate and, consequently, biological competitiveness.

Toward a Whole-Cell Model of Human Embryonic Stem Cells: A Composable Model of Metabolism Built on a Scalable Framework for Integrating Multiple Pathway Submodels

Authors:

Chew, Y. H., Icahn School of Medicine at Mount Sinai
Karr, J. R., Icahn School of Medicine at Mount Sinai
Sundstrom, A., Icahn School of Medicine at Mount Sinai

Stem cell behaviors such as self-renewal and differentiation result from complex interactions among signaling, gene regulation, metabolism, and other pathways. To understand how these pathways collectively determine cell behavior, we aim to develop the first whole-cell computational model of a human cell, focusing on the well-characterized H1 human embryonic stem cell (hESC) line.
Our approach is to build and integrate submodels of each pathway. To begin, we have developed a scalable framework for integrating multiple pathway submodels and built a genome-scale submodel of single-cell metabolism.

The integration framework consists of state variables that represent the instantaneous configuration of a cell such as its mass, volume, and composition including the abundance and compartmental location of each molecular component. We developed the initial conditions for the framework by integrating extensive data including the DNA sequence of H1-hESCs; the location of each transcript, exon, and protein lifted from the hg19 reference; protein-complex compositions; macromolecular contents; small-metabolite concentrations; RNA expression; protein abundances; and the stoichiometry, catalysis, rate law, and kinetic data of each reaction.

We built the metabolism submodel by: (1) expanding Recon 2.2 to produce and recycle the metabolites that are needed and generated by other pathways, respectively; (2) refining the submodel for integration into a stochastic dynamical model of individual cells; and (3) using multi-omics and biochemical data to contextualize the submodel to represent H1-hESCs. The metabolism submodel correctly predicts the essentiality of 97% of metabolism-associated genes in hESCs. 87.5% of the false negative have non-metabolic functions, highlighting the need for integrated models that describe multiple pathways.

Going forward, we will create submodels of other pathways, integrate the submodels, and use the integrated model to investigate stem cell self-renewal. Ultimately, we believe whole-cell models have the potential to transform basic science and medicine through unprecedented*in silico* experiments personalized to each patient.

Towards a Genome-Wide Transport Systems Encoding Genes Tracker

Authors:

Dias, O., University of Minho
Lagoa, D., University of Minho
Ferreira, E., University of Minho

The Transport Systems Tracker (*TranSyT*) is a tool to identify transport systems and the compounds carried across membranes by those systems, annotating the proteins with the Transporter Classification Database (TCDB)[1] families numbers. In addition, this tool also generates the respective transport reactions.

TranSyTâ€™s internal database contains every possible reaction for all compounds described in each TCDB entry, as well as their hierarchical children metabolites.

TranSyT starts by assigning identifiers to metabolites, as TCDB does not provide cross links to any database. These metabolites are identified by searching a BioDB, a database developed in-house, for all possible names and synonyms. The relationships between compounds (hierarchical ontology) was also determined using BioDB, which combines information retrieved from several sources such as ModelSEED, KEGG, MetaCyc and BiGG.

The second step involves retrieving the TC families information from TCDB, which allows to find the suitable transport type for each metabolite (e.g. symport and antiport) and direction (in/out or out/in).

All generated information is then stored in a Neo4j graph database, which allows *TranSyT* to retrieve the required reactions for the genome in study, saving time and resources.
All sequences of the genome in study are aligned against the whole set of records retrieved from TCDB, containing the sequences of well-known transport systems, using BLAST+.

After the alignments, only genes having evidences of encoding transport systems are associated to transport reactions. The compartmentalization of such reactions is also possible, as the direction of the reaction was forehand defined.

TranSyt was developed in Java and is a the next iteration of the previously developed TRIAGE[2] tool.

References:

1 Saier, M. H. *et al. Nucleic Acids Res.*34, D181â€"D186 (2006)
2 Dias, O. *et al. IEEE/ACM Trans. Comput. Biol. Bioinforma.* XX, 1â€"1 (2016)

Towards a Metabolic and Expression Model of the Metabolically Versatile Bacterium Rhodopseudomonas Palustris

Authors:

Alsiyabi, A., University of Nebraska - Lincoln
Immethun, C., University of Nebraska
Saha, R., University of Nebraska-Lincoln

Rhodopseudomonas palustris is a metabolically versatile Purple Non-Sulfur Bacterium (PNSB). Depending on growth conditions, *R. palustris* can operate in either one of four different forms of metabolism: photoautotrophic, photoheterotrophic, chemoautotrophic, and chemoheterotrophic.*R. palustris* is also a facultative anaerobe and is capable of fixing nitrogen. This metabolic flexibility suggests that *R. palustris* would be an ideal model organism for studying the network of interacting reactions, and how itâ€™s altered in response to changing conditions. Moreover, exploration into *R. palustrisâ€™* industrial applications has just begun with research being generated on the production of hydrogen, methane, and polyhydroxybutyrate (PHB). This study aims to combine systems and synthetic biology by reconstructing *R. palustris*â€™ metabolic and expression (ME-) network and establishing a constitutive promoter library to gain a mechanistic understanding of the organismâ€™s functional capabilities under different conditions.

In this study, a genome-scale metabolic model was first constructed following a well-developed framework established in the lab. The metabolic model currently contains 925 reactions, 1220 metabolites, and 1134 genes. An expression matrix (E-matix), containing transcription, translation, and post-transcriptional modifications, is currently being reconstructed. After the reconstruction of this matrix, the two networks will be integrated and previously established â€˜coupling constraintsâ€™ will be implemented to link the synthesis of macromolecular molecules to catalysis of metabolic components. Moreover, preliminary experimental results showed that the bacterium is able to utilize a host of different carbon sources, including a host of lignin-derived compounds. An IPTG induced promoter was also established to express the CRISPRi components. This system will be used to verify and iteratively improve model predictions, resulting in a well-established mechanistic understanding of the organismâ€™s cellular functions. The predictive power of this work will expand as more omics datasets become available, facilitating the establishment of this metabolically versatile, non-model microorganism as a biotechnology platform.

Untangling Dynamic Protein Behaviour with Rule-Based Modelling and Sensitivity Analysis

Authors:

Denhardt-Eriksson, R., EPFL - LCSB

The endoplasmic reticulum (ER) is key to synthesizing mature, functional proteins. Its complex structure consisting of large thin sheets densely packed around the nucleus and elongated tubules reaching throughout the cell is key to its function, and is dynamically regulated in response to cellular signals. CLIMP-63 is the only known protein to play a role in shaping the ER, while also being the target of palmitoylation, one of the few reversible post-translational lipid modifications. In addition to its role in ER structure regulation, CLIMP-63 plays a role as a receptor at the cell surface, where it is known to bind to several ligands. These multiple roles make untangling the regulation of CLIMP-63 abundance and localisation a non-trivial task. We employed rule-based modelling to build a dynamical representation of CLIMP-63 birth, dimerisation, palmitoylation, traficking and degradation. Experimental data in the form of time-dependent radioactive labelling was used to calibrate the model with heuristic optimisation methods. Multiple predictions and hypothesis were generated as to the regulation of CLIMP-63 levels and localisation. Sensitivity Analysis was used on the calibrated model to maximise the extraction of useful knowledge and hypotheses.

Use of a Symbiotic Genome-Scale Model for the Study of Wolbachia's Pathogen Blocking Effect in Zika Replication

Authors:

Jiménez, N. E., Centre for Biotechnology and Bioengineering
Gerdtzen, Z. P., Centre for Biotechnology and Bioengineering
Salgado, J. C., Centre for Biotechnology and Bioengineering CeBiB
Olivera-Nappa, A., Centre for Biotechnology and Bioengineering CeBiB
Conca, C., Center for Mathematical Modeling

Growth rate (Âµ) and yield (Y) are fundamental features of microbial growth. However, we lack a mechanistic and quantitative understanding of the Âµâ€“Y relationship. Studies pairing computational predictions with experiments have shown the importance of maintenance energy and proteome allocation in explaining rate-yield tradeoffs and overflow metabolism. Recently, adaptive evolution experiments of*Escherichia coli* reveal a phenotypic diversity beyond what has been explained using a simple Âµâ€“Y relationship. Here, we identify a two-dimensional Âµâ€“Y tradeoff in adapted *E. coli* strains where the dimensions are (A) a tradeoff between Âµ and Y and (B) a tradeoff between substrate uptake rate (q_{glc}) and Y. We employ a multi-scale modeling approach, combining a previously reported small-scale proteome allocation model with a genome-scale model of metabolism and gene expression (ME-model), to develop a quantitative description of this two dimensional Âµâ€“Y relationship using data for *E. coli* K-12 MG1655. The analysis provides a mechanistic explanation of the two-dimensional Âµâ€“Y tradeoff. Furthermore, the analysis identifies modifications to the P/O ratio as a potential mechanism that enables the q_{glc}-Y tradeoff. Thus, the Âµâ€“Y tradeoffs that govern microbial adaptation to new environments are more complex than previously reported, and they can be understood in mechanistic detail using a multi-scale modeling approach.

Utilizing RNA-Seq Data in Bayesian Estimation of Gene Activity States

Authors:

Sorensen, E., Pacific Lutheran University
Boehme, J., Oregon State University
Gasdaska, A., Emory University
DeJongh, M., Hope College
Best, A., Hope College
Lindsey, W., Dordt College
Tintle, N., Dordt College

Recently, there has been interest in exploring how to infer gene activity states (e.g., whether a gene is active or inactive in a particular condition of interest) from genome-wide transcriptomic data. This knowledge is useful in many downstream applications, including the potentially improved use of transcriptomics data to improve flux predictions in metabolic models. Recently, a rigorous Bayesian approach (*MultiMM*) to classifying gene activity states was proposed that leverages a priori knowledge of operon structure as well as genome-wide transcriptomics data from multiple conditions in order to classify gene activity states. However, the *MultiMM*approach was developed for use on microarrays, and only evaluated on a very large set of over 900 *E. coli* arrays. Here, we extend the Bayesian model to RNA sequencing data and then evaluate its performance. Importantly, we evaluate performance in situations with both large (100s) and small (10s) of conditions, and provide intuition on necessary sample sizes for robust performance.

Visualizing Metabolic Network Dynamics through Time-Series Metabolomics Data

Authors:

Dräger, A., Center for Bioinformatics Tübingen (ZBIT)
Buchweiz, L. F., University of Tübingen
Yurkovich, J. T., University of California, San Diego
Blessing, C. M., University of Tübingen
Kohler, V., University of Tübingen
Schwarzkopf, F., yWorks GmbH
King, Z. A., University of California, San Diego
Yang, L., University of California, San Diego
Jóhannsson, F., University of Iceland
Sigurjónsson, Ó., Reykjavik University
Rolfsson, Ó., University Of Iceland
Heinrich, J., University of Tübingen

Lactic acid bacteria (LAB) have long been used by humans in the food industry and is*Lactobacillus reuteri* one of them. Additionally, *L. reuteri* is also known to produce several different compounds, with potential biotechnological interest, such as; the antimicrobial compound reuterin, vitamin B12 and 1,3-propanediol, which can be used as a starting material for plastic production. In an attempt to further explore*L. reuteri* as a cell factory for production of valuable compounds, a genome-scale metabolic model was reconstructed and thoroughly curated and validated using data from the literature and several experimental data sets, including knockout strains. The biomass reaction was built from species specific data when available and alternatively from *L. plantarum* and *L. lactis*.

L. reuteri metabolizes glucose via the phosphoketolase pathway. Genome data indicates the absence of a phosphofruktokinase (PFK), which results in an inactive EMP pathway. Model simulations show that PFK does not provide a significant advantage in terms of growth.

A pathway prediction algorithm was used to identify several heterologous pathways for production of 1-propanol. Theoretical yields of 1-propanol were evaluated for multiple carbon sources. The most promising pathways were further optimized *in silico* by identifying targets for overexpression and knockouts for increased production of 1-propanol.

AUTHOR INDEX

AUTHOR INDEX

AUTHOR INDEX

AUTHOR INDEX

AUTHOR INDEX

AIChE
120 Wall Street, FL 23
New York, NY 10005-4020

ISBN 978-1-5108-7587-6

9 781510 875876

CURRAN ASSOCIATES INC.
proceedings
.com